The
Science Times
Book of
BIRDS

The Science Times Book of BIRDS

EDITED BY NICHOLAS WADE

THE LYONS PRESS

Printed in the United States of America

Designed by Joel Friedlander Publishing Services, San Rafael, CA

10 9 8 7 6 5 4 3 2 1

Library of Congress Cataloging-in-Publication Data

The Science times book of birds / edited by Nicholas Wade.
     p. cm.
   ISBN 1-55821-605-7 (cloth)
    1. Birds. I. Wade, Nicholas. II. Science times.
QL676.S38   1997
598—dc21   97-14718
                                    CIP

Contents

9   Birds and Birders

Introduction

IF BIRDS HAVE A SPECIAL PLACE in people's hearts above that of most other wild animals, it is surely because they are such obvious miracles of design.

From the lightweight bones to the powerful heart, every feature of a bird's body is adapted to the exigencies of flight. Liberated into the medium of air, these creatures are nature's free spirits, metaphors for grace and rhapsody and all the virtues that flow from having learned how to defy the enchaining hand of gravity.

From admiring birds to studying them for their own sake is a small step that many have taken. Much of modern biology has been shaped by what has been learned from birds, from Darwin's discovery of the evolution of the Galapagos finches to current studies by neuroscientists of how birds grow new brain cells to learn certain tasks.

Reflecting the keen ornithological interests of both the public and scientists, the Science Times section of *The New York Times* has carried many articles on birds. Between 1990 and 1996, while I was the section's editor, new discoveries about birds seem to have been particularly prolific.

Newly unearthed fossils have advanced understanding of the evolution of birds, even though their precise relationship with dinosaurs remains unclear. The family life of birds has turned out to be more akin to soap opera than suspected. Contrary to the belief that many bird species mate for life, a new survey has found that some 30 percent of the chicks in any given nest tend to be sired by a father other than the resident male.

Well studied as birds are, they continue to produce surprises. The Seychelles warbler can apparently determine the sex of its offspring, producing different ratios of males to females depending on environmental conditions. The barn swallow's curious preference that her mate have a symmetrical tail turns out to be no mere caprice; symmetry measures the male's relative freedom from parasites, and hence his general state of health.

Biologists are gaining new respect for birds' mental abilities. Particularly striking is the recent discovery that a species of crow in the forests of New Caledonia has a tool-using ability that seems comparable to that of early humans.

In the course of their long evolution, birds have accumulated a rich repertoire of behaviors. Biologists have a sense of urgency in their work because so many birds are threatened by the encroachments of civilization.

Tropical forests are being cleared in a rush for timber or agricultural land. Even in the United States, where farmland is reverting back to forest, songbirds are in steady decline. Though the reason was thought to be deforestation of the birds' winter homes in the Caribbean and Central America, it now seems that the fragmentation of woodlands in the United States is responsible, since predators dependent on man, like cats, raccoons and squirrels, then gain access to the birds' nests.

Grassland species are no better off. Grassland often makes good agricultural land and as it is cleared many species of shrike, a key grassland bird, have declined throughout the world.

Captive breeding programs have sometimes succeeded in rescuing endangered birds from the brink of extinction, as in the notable case of the peregrine falcon. But such programs do not always work, and in any case are too expensive to mount except on behalf of a handful of popular species, such as the California condor.

The articles in this collection are intended to bring readers abreast of the latest research findings in bird biology and behavior. They convey the richness and scope of present day ornithology, as well as the importance of birds in helping to understand the natural world around us.

Many of these articles appeared on the environment page of Science Times, edited by William A. Dicke. The idea for the book came from Lilly Golden of Lyons & Burford. To the writers of Science Times belongs the credit for whatever amusement or instruction the reader may find.

—NICHOLAS WADE, Fall 1997

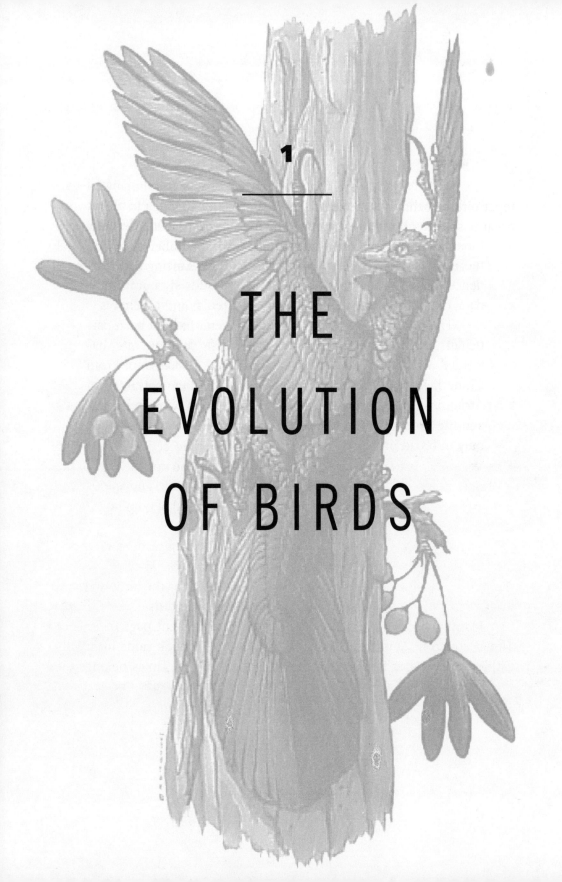

1

THE EVOLUTION OF BIRDS

There are many bird lovers, and many people fascinated with dinosaurs, and not a little willingness to believe that one evolved from the other. And maybe that was the case. But the precise evolutionary pathway by which dinosaurs morphed into birds is still uncertain. So fragmentary is the evidence that each new fossil of an early bird adds significant new data, often interpreted differently by adherents of rival schools.

The recent crop of early bird fossils seems to have been particularly rich. The Mongolian desert, a classic dinosaur site, has yielded a flightless, bird-like creature named *Mononychus*. From China has come *Confuciusornis sanctus,* and a new specimen of *Archaeopteryx* has come to light. But the origins of avian lineage and the events that molded scale into feather can still be traced only in broad brush stroke.

Feathered Dinosaur or Flightless Bird: A New Find Stirs the Dust

THREE BIRD WATCHERS of an unusual kind, without binoculars or field guides by Roger Tory Peterson, leaned over the laboratory table and compared pieces of ancient evidence central to one of the most controversial issues in paleontology today: the history of avian evolution and early flight.

These paleontologists at the American Museum of Natural History in New York City have ruffled scholarly feathers with the announcement of their discovery of a strange new type of dinosaur in the Gobi Desert of Mongolia. The 75-million-year-old fossil animal, about the size of a turkey, was actually a flightless bird, they contend, and it perched firmly on the evolutionary tree as a transitional figure between certain carnivorous dinosaurs and modern birds.

Such an interpretation had the effect of a setter's flushing a covey of quail into flight over hunters with loaded shotguns at the ready.

Bird it never was, fired the ornithologists, renewing with even greater heat the longstanding debate over how closely related are dinosaurs and birds. The theory popular among paleontologists is that birds are the direct descendants of dinosaurs. Most ornithologists disagree, contending that birds arose from reptiles only distantly related to dinosaurs. On both sides, the argument has focused on supposed anatomical similarities between dinosaurs and birds and whether flight could have begun with dinosaurs from the ground up or with protobirds from the trees down.

Many dinosaur experts, while agreeing that this was an important and surprising find, questioned several assumptions being made to support the conclusion that the newly discovered animal is a flightless bird. They also had strong reservations about its supposed place in bird evolution.

5

Michael Rothman

If no cease-fire seemed imminent, there was at least the feeling of hopeful excitement lingering after the smoky fusillade, as expressed by Dr. Paul Sereno, a dinosaur paleontologist at the University of Chicago. "We're just bristling with new ideas about birds in the Mesozoic," he said, referring to the era sometimes known as the Age of Reptiles, from 245 million to 65 million years ago.

So it was that the three paleontologists who believe they have an early bird in the hand sought to explain and defend themselves. Spread out on the table at the museum were casts of the remains of three of the most famous ancient fossil birds, as well as their own puzzling find, which they have named *Mononychus olecranus*.

"This is the evidence of early bird evolution," said Dr. Mark A. Norell, assistant curator of vertebrate paleontology at the museum and a leader of the discovery team. He pointed first to the impression of the earliest established bird-like fossil. This was *Archaeopteryx*, etched in limestone and found in Bavaria 130 years ago.

With the teeth, long neck and tail of a dinosaur but the hollow bones and feathered wing of a bird, *Archaeopteryx* is usually described as a dinosaur on the way to becoming a bird, if not a bird already. Judging by its asymmetrical feather pattern, similar to that of modern birds, it probably could fly in some fashion. Six well-preserved specimens have been found in the 150-million-year-old limestone, lithified sediments from a prehistoric lagoon, and a seventh discovery is expected to be reported soon by German scientists.

"If this is the most primitive of all birds, " Dr. Norell said, nodding at *Archaeopteryx*, "then these are the sec-

Transformations needed for flight
Powerful wing-flapping muscles enable modern-day birds to fly. They attach to the wishbone and sternum in the breast. *Archaeopteryx* had a small wishbone but no sternum, which would have greatly limited its ability to fly.

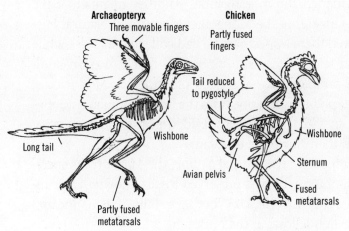

Archaeopteryx
Three movable fingers

Chicken
Partly fused fingers

Tail reduced to pygostyle

Wishbone

Long tail

Wishbone

Avian pelvis

Sternum

Fused metatarsals

Partly fused metatarsals

Michael Rothman

ond and third most primitive ones we know of." He picked up casts bearing the impressions of two recently discovered sparrow-size specimens that scientists agree were flying birds. One was the 135-million-year-old *Sinornis* from China, and the other, slightly younger one was *Iberomesornis* from Spain.

Sinornis had a keeled breast bone, similar to the one to which flight muscles are attached in modern birds. Both species had an elongated strut-like pectoral bone, the coracoid, showing that the musculature of their breasts worked to elevate the wing much as it does in today's birds. Both had given up the long tails of their reptilian heritage and were left with only a pygostyle, the "pope's nose," a stubby vestigial tail of fused vertebrae out of which grew a feather fan essential to flight maneuvering.

Finally, Dr. Norell turned to *Mononychus* and, anticipating the question of how such a flightless, wingless animal living some 50 million years later could fit comfortably into any picture of avian evolution, said: "Whatever you would expect, it wouldn't be something like this. You would expect a fully flying bird."

The *Mononychus* fossils were found in the Gobi, where the American Museum and the Mongolian Academy of Science are conducting a four-year expedition in search of dinosaurs and other remains of prehistoric life. As often happens, the scientists returning from last year's season did not know what they had until, back in the laboratory, they examined the bones of a thrush-size animal. This was, as they would later understand, a juvenile *Mononychus*.

Dr. Norell then widened the investigation. Another member of the expedition, Dr. Perle Altangerel of the Mongolian Museum of Natural History in Ulan Bator, showed them a specimen he had collected in 1987 while working with a Russian team. Close study revealed that it was an adult *Mononychus*. Two other samples have also been identified as *Mononychus*, another from the 1992 expedition and one from the Roy Chapman Andrews expedition to the Gobi in 1923, which had resided in the museum since then with no other label than "bird-like dinosaur."

In the first formal report, in the journal *Nature,* the *Mononychus* specimens were described as "among the best preserved fossils known of a primitive bird." The authors were Dr. Perle, Dr. Norell, Dr. Luis M. Chiappe and Dr. James M. Clark, research fellows at the museum.

Mononychus shares several anatomical traits with modern birds that *Archaeopteryx* does not, including its keeled breast bone, fused wrist bones and aspects of its long leg bones. Its braincase is especially bird-like. In contrast to other primitive bird-like fossils, which are two-dimensional impressions on stone, the *Mononychus* bones are preserved in three dimensions and can be held next to the bones of dinosaurs for detailed comparisons of relationships.

But *Mononychus* does not have wings. Each of its two forelimbs is only three inches long and ends in a single stout claw. *Mononychus* means "one claw." Both the discoverers and their critics suggest that the animal used these strong forelimbs and claws for digging. Clearly, *Mononychus* did not fly.

"Just because it doesn't fly doesn't mean it's not a bird," Dr. Clark argued.

The fossils raise two possible interpretations of avian evolution, as Dr. Norell said. Flight could have evolved very early in the bird lineage, before *Archaeopteryx,* and continued to modern birds—with the exception of the line that led off to *Mononychus,* where it was lost. Or if the common ancestor of *Archaeopteryx* and all other birds was flightless, that would mean that flight evolved independently in one lineage that included *Archaeopteryx* and in another that included the flightless *Mononychus,* where it evolved later.

"Our point is that both hypotheses are equally simple, requiring two evolutionary events," Dr. Norell explained. "Neither can be preferred on the basis of the evidence at hand."

The grounded kiwi, emu, ostrich and rails, all of which had flighted ancestors, attest to the fact that flying species can evolve into flightless ones. But paleontologists like Dr. Sereno questioned whether the new discovery could support a hypothesis that flight was invented more than once in the mainstream of avian evolution. Dr. Alan Feduccia, an ornithologist at the University of North Carolina at Chapel Hill, was more emphatic. "It's just replete with problems," he said.

In any event, scientists pointed out, the fossil record includes several flying species in the millions of years after *Archaeopteryx* and before *Mononychus.* By the time of *Mononychus,* moreover, precursors to today's gulls, plovers and pelicans were already flying along ancient shores.

Dr. John H. Ostrom, a paleontologist at Yale University and authority on *Archaeopteryx,* said the discoverers of *Mononychus* were making "some very large but reasonable assumptions." But he expressed reservations about their inferences based on the keeled breast bone.

"A keeled breast bone does not in and of itself support a flight hypothesis," he said. "And none of the rest of the animal seems to fit the hypothesis."

Ornithologists focused on the possibility that this was only a bird-like dinosaur, not a line of dinosaurs evolving into birds. They said that not enough attention was being paid to the possibility that the resemblances were examples of what is known as divergent evolution, in which animals of two lineages independent of any common ancestor, like bats and birds, develop similarities in adapting to the same kinds of ecological niches.

Dr. Storrs Olson, curator of birds at the Smithsonian Institution in Washington, who examined the *Mononychus* fossils, said: "Some of the leg bones are superficially quite bird-like. But this was very late in dinosaur history. Many dinosaurs were becoming more and more bird-like by that time. But that does not mean they were birds and does not establish a close bird-dinosaur relationship."

Indeed, the keeled breast bone in *Mononychus* could be an argument against its birdness. Dr. Feduccia, author of the definitive *The Age of Birds,* noted that ostriches, kiwis and all other known flightless birds lost their essential flight apparatus, like the keeled breast bone, after they became flightless. He suggested that the breast bone in the case of *Mononychus* was developed to support muscles for digging, not flight.

Interpretations and criticism of *Mononychus* reflect larger disputes over the course of avian evolution. Many paleontologists, especially dinosaur experts, contend that birds evolved directly from a group of agile bipedal dinosaurs known as theropods. It was a comparison of the anatomies of *Deinonychus,* one of these theropods, and *Archaeopteryx* by Dr. Ostrom in the 1970's that seemed to clinch the dinosaur-bird link.

No one disputes that birds evolved from reptiles, but ornithologists generally hold that the transition occurred in a reptilian line that split off long before the rise of dinosaurs. Some paleontologists insist that *Archaeopteryx* was a ground-dwelling dinosaur with feathers and wings. Writing in the journal *Evolution,* Dr. J. R. Speakman, a zoologist at the University of Aberdeen in Scotland, observed that the mass of flight muscles in *Archaeopteryx* was insufficient for it to have sustained powered flight.

But Dr. Feduccia recently concluded from an analysis of the *Archaeopteryx*'s claws that it was an almost fully evolved bird that perched in trees, was considerably advanced aerodynamically and was probably capable of powered flight to some degree.

Paleontologists and ornithologists are also sharply divided on the question of how birds or protobirds first mastered that most graceful and envied of their talents: flight.

On this issue, many paleontologists think that flight developed from the ground up. Creatures like *Archaeopteryx* ran, jumped and flapped their feathered wings, perhaps to catch insects, and they eventually found themselves ascending in flight. The alternative is that flight began from the trees down, an idea favored by ornithologists.

For Dr. Feduccia this is where the dinosaur-bird connection goes into a tailspin. Beginning flight from the ground up with only rudimentary flight muscles, he said, is a "biophysical impossibility." And why did some dinosaurs have feathers in the first place? Paleontologists surmise that feathers, the quintessential feature of birdness, evolved from dinosaur scales as an insulating cover, but Dr. Feduccia said this was probably unnecessary.

"To clothe a ground-dwelling, warm-blooded reptile with feathers for insulation is tantamount to insulating an ice truck with heat shields from the space shuttle," he said.

The most reasonable hypothesis, Dr. Feduccia said, is that feathers evolved for feather-assisted jumping from limb to limb by tree-dwelling reptiles, the protobirds, which were not dinosaurs but a separate lineage with only distant relationship from some common ancestor in the past.

Still, the idea that birds are the living descendants of dinosaurs, an appealing one to dinosaur aficionados, has been the prevailing view, and the new research could be seen as both supporting and complicating this picture. Dr. Norell thus finds himself in the middle of the expanded dispute. Looking up from the laboratory table, he squared his shoulders, as if to confront the critics, and asserted, "It's a bird!"

—JOHN NOBLE WILFORD, April 1993

Two Clues Supporting Idea That Birds Arose from Dinosaurs

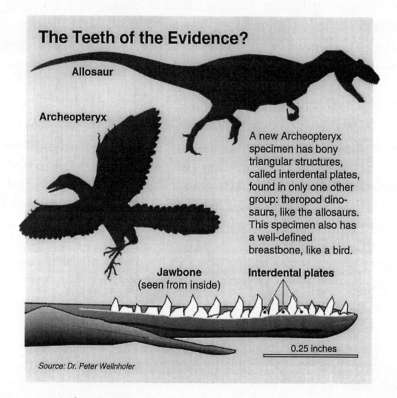

The Teeth of the Evidence?

Allosaur

Archeopteryx

A new Archeopteryx specimen has bony triangular structures, called interdental plates, found in only one other group: theropod dinosaurs, like the allosaurs. This specimen also has a well-defined breastbone, like a bird.

Jawbone (seen from inside)

Interdental plates

0.25 inches

Source: Dr. Peter Wellnhofer

FROM THE JAW of an ancient bird unearthed in a German slate quarry and from fossil leg bones of baby dinosaurs found in Montana, two groups of scientists have turned up striking new evidence that birds are closely related to dinosaurs.

For more than a century, paleontologists have noticed similarities between the fossils of many dinosaurs and the skeletal structure of modern birds, and many scientists say they believe that birds evolved directly from dinosaurs. Some ornithologists, on the other hand, argue that these similarities are coincidental, resulting from parallel but independent development of birds and dinosaurs. The debate remains far from settled.

Against this background, Dr. Peter Wellnhofer of the Munich Museum in Germany and Dr. John Ostrom of Yale University report that they have found a surprising feature in a recently discovered specimen of the ancient bird *Archaeopteryx* that seems to back the paleontologists' position. The feature they noticed is found in only one other group of animals, extinct or modern: the theropod ("beast-footed") dinosaurs. Most of the theropods, typified by the big Jurassic hunters called allosaurs, were active, bipedal flesh-eaters.

The feature, found on the inner side of the *Archaeopteryx*'s lower jaw at the gum line, is a row of small, triangular, bony plates, each positioned between the roots of the bird's teeth.

Dr. Wellnhofer reports the feature in the German journal *Archaeopteryx,* and Dr. Ostrom discussed it in an interview.

Dr. Ostrom said he and Dr. Wellnhofer had independently spotted the peculiar *Archaeopteryx* feature soon after the specimen was sent to the Munich Museum a year ago. The fossilized Jurassic-period bird, which lived 150 million years ago, is similar but not identical to six previously discovered *Archaeopteryx* specimens, which are regarded as the most valuable of all fossils. The latest fossil was discovered in August 1992, but the scientists studying it have only now concluded their study and reported their findings.

Some paleontologists have said all along that the *Archaeopteryx* should be regarded as a feathered theropod dinosaur, which may or may not have been able to fly. The absence of bony breast bones in the previously discovered six archeopteryxes led some researchers to believe that these animals could not fly.

But the latest specimen, Dr. Ostrom said, has a well-defined breast bone. Because of this and some other distinctions, he said, Dr. Wellnhofer has proposed a new species name for it: *Archaeopteryx bavarica*. The other six specimens have the species name *Archaeopteryx lithographica.*

"This *Archaeopteryx* is going to be a real bombshell for paleontology," Dr. Ostrom said.

Another piece of evidence for a link between dinosaurs and birds comes from investigators at the University of Wisconsin and the Museum of the Rockies at Bozeman, Montana, who disclosed the results of their study of leg bones from young duckbill dinosaurs called maiasaurs. Some paleontologists say they believe that members of the species cared for their young for years after hatching. Many paleontologists say the nurturing behavior of maiasaurs and certain other dinosaur species sets them apart from modern reptiles, in which such behavior is supposedly rare.

The report on the maiasaurs, published in the journal *Science,* adds fuel to the debate over the origin of birds. Even before its publication, the paper was drawing criticism.

Claudia Barreto, a graduate student at the University of Wisconsin School of Veterinary Medicine and the lead investigator of the group issuing the *Science* report, explains that the leg bones of young maiasaur dinosaurs, like those of all young animals, grew by adding cartilage to their ends and then replacing the cartilage with bone.

Ms. Barreto said that microscopic examination of juvenile maiasaur leg bones revealed that their growth plates—the junctions between the rigid bone and the cartilage it replaces—were preserved as fossils. These junctions, the scientists found, were wavy in shape, making them similar to the growth plates found in the bones of young birds but unlike the growth plates of modern reptiles or mammals, which form straight junctions at the region where cartilage is converted into bone.

Ms. Barreto and her colleagues also found that fossilized dinosaur chondrocytes, or cartilage cells, are shorter and rounder than are the corresponding cells in mammals and reptiles, but similar to those of birds. An examination using a scanning electron microscope also showed a pattern of calcification of growing dinosaur cells that more nearly matches that of birds than reptiles and mammals.

"What we've done," Ms. Barreto said, "is identify another shared characteristic of birds and dinosaurs."

Many paleontologists, including Dr. Ostrom at Yale, Dr. Robert T. Bakker of the University of Colorado and John R. Horner of the Museum of the Rockies, have long contended that at least some dinosaurs were warm-blooded, that many of them were active creatures and that modern birds, whose skeletons closely resemble those of some dinosaurs, evolved from dinosaurs.

Ms. Barreto said these views were strongly reinforced by the bones she and her colleagues studied, which belonged to maiasaur fossils excavated by Mr. Horner from the Two Medicine rock formation in Montana.

Two decades ago, Mr. Horner, the discoverer of the maiasaur species of duckbills, unearthed evidence that maiasaur mothers looked after their young in nests, caring for the chicks for years after they had hatched. This and other hints of the behavior of these 30-foot-long dinosaurs suggest to some scientists that they behaved more like birds than modern reptiles.

Although most paleontologists have come to accept the theory of a dinosaur ancestry for birds, some scientists strongly disagree. Among them is Dr. Alan Feduccia at the University of North Carolina at Chapel Hill, who contends that the similarities in dinosaur and bird skeletons are coincidental, representing parallel but independent lines of evolution.

"I just don't see that maiasaurs and other dinosaurs that lived toward the end of the Cretaceous period could have anything to do with the ancestry of birds," he said in an interview. "Some 80 million years earlier, before we see any of these supposed bird-like features in Cretaceous dinosaurs, *Archaeopteryx* was happily flying around—a true bird, which existed long before these late Cretaceous dinosaurs. I see the similarities as the product of convergent evolution."

Dr. Feduccia also rejects the argument that maiasaurs resembled birds in their care of young. "Crocodile babies also stay around their nests for up to four years," he said. "Parents guard them, and the babies ride around on their mothers' backs. But crocodiles are not birds."

But Dr. Ostrom maintains that the latest *Archaeopteryx* discovery in Germany will swing the balance of scientific opinion even more strongly to the view that this animal was closely related to the common ancestor of later dinosaurs and modern birds.

The new *Archaeopteryx* fossil is only the seventh *Archaeopteryx* specimen ever found. All seven were discovered in slate quarries in Solnhofen, and they are so rare and precious to scientists that a single *Archaeopteryx* fossil has been valued at $5 million. One famous *Archaeopteryx* fossil, owned by its discoverer, Eduard Opitsch, disappeared when he committed suicide in 1991, and it is presumed to have been stolen from his estate.

—MALCOLM W. BROWNE, December 1993

Very Early Bird Had a Way to Catch Worms: In a Beak

LONG AGO, when hens' teeth were anything but scarce, there lived a bird by a lake surrounded by lush forest. This bird, about the size of a bantam rooster, had a long tail like its reptilian ancestors and vestigial forearms that ended in claws, handy for climbing trees. It also flapped its feathered wings in flight over a late Jurassic landscape dominated by reptiles, including dinosaurs.

In one striking respect, this primitive bird had taken a precocious evolutionary step that seemed to anticipate every bird of today, every eagle and robin and chickadee: there was not a tooth in its head. This is the earliest bird known to paleontology to have abandoned the toothy jaws of its reptilian ancestors, replacing them with a true avian beak.

Fossil remains of this bird were found in China in 1994, and paleontologists, astonished and excited by the discovery, say the findings could have revolutionary effects on thinking about bird evolution.

The exact age of the bird is not known, but paleontologists say it was probably close in time to *Archaeopteryx,* the transitional reptile-bird that lived about 147 million years ago in what is now Germany and is recognized as the oldest known bird. If that is the case, the new fossil species lived 70 million years earlier than the previously oldest known toothless bird, *Gobipteryx,* from Mongolia.

The discovery of a skull, wing, two feathered legs and a pelvis in ancient lake-bed sediments was made by a farmer in the Liaoning province in northeastern China, near the border with North Korea. Chinese and American paleontologists described the fossils and discussed their implications in the journal *Nature.* Their name for the new fossil species is *Confuciusornis sanctus,* for holy Confucius bird.

Michael Rothman

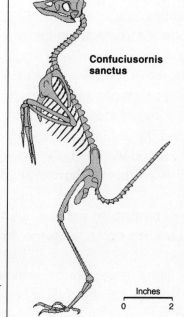

Confuciusornis sanctus

John Papasian

Inches
0   2

Artist's reconstruction of Confuciusornis sanctus, *the earliest known bird with a beak, climbing a Ginkgo yimaenis tree.*

The Confucius bird provides compelling evidence that nature's initial experimentation with birds must have spread quickly into a global phenomenon played out in different habitats and marked by seemingly rapid evolutionary transitions, even some false starts.

Indeed, the researchers suggest that the Confucius bird occupied a separate limb of the avian family tree that branched off soon after the emergence of *Archaeopteryx,* leading to *Gobipteryx* and eventual extinction. These birds somehow lost their teeth and developed horny bills, an adaptation that the main line of avian evolution did not exhibit until the end of the Cretaceous period, about 65 million years ago.

For almost a century *Archaeopteryx* has been alone on its perch as the early bird of the Jurassic geological period. But the new findings suggest that birds in several forms and stages of evolution probably existed at the time, or shortly thereafter.

Since the Confucius bird was found on the other side of the world from *Archaeopteryx,* paleontologists said, the discovery shows that birds in different forms were a widely dispersed phenomenon. They had even adapted to different habitats: the Chinese fossils were found in a freshwater environment; *Archaeopteryx* came from marine lagoons.

"We now have birds that were different from *Archaeopteryx,* very different," said Dr. Larry D. Martin, a paleontologist at the University of Kansas at Lawrence and one of the authors of the *Nature* report. "We also know there was some diversity in birds at that time, both in geography and in design."

An analysis of the fossils showed that the wing skeleton, including the long fingers and big claws, and two legs retained the primitive, almost reptilian features found in *Archaeopteryx.* But the skull, with its beak, represented a dramatic innovation. The horny bill is assumed to have evolved from reptilian scales.

The Confucius bird shows another sign of modernity: the first direct evidence of body feathers. The only preserved feathers on *Archaeopteryx* are on its wings.

Another author, Dr. Alan Feduccia of the University of North Carolina at Chapel Hill, said the Confucius bird's "transition to a modern avian beak so swiftly is really astounding."

In their report, Dr. Martin, Dr. Feduccia and their Chinese colleagues, Dr. Lian-hai Hou of the Institute of Vertebrate Paleontology in Beijing and

Zhonghe Zhou, a graduate student at Kansas, say the findings hint at discoveries yet to be made at this critical stage in bird evolution. They write: "These specimens provide evidence for either an undiscovered pre-*Archaeopteryx* or a rapid post-*Archaeopteryx* evolution in birds."

Other paleontologists familiar with the fossils quibble with the suggestion that the Confucius bird lived as early as the Jurassic period, and the authors of the *Nature* report acknowledge that the geology of the region made it difficult to pin down the timing.

But the scientists echo the reaction of Dr. Mark Norell of the American Museum of Natural History in New York City: "It's a very, very important specimen." Dr. Luis Chiappe, a research associate at the museum who has examined the fossils, said they would yield valuable information about the early evolution of birds and made him all the more curious to discover the common ancestor of *Archaeopteryx* and *Confuciusornis*.

But even members of the research team do not quite agree why an early bird might cast off its reptilian teeth.

Dr. Feduccia says it was one more weight-saving modification to aid flight; this is the conventional explanation.

But Dr. Martin doubts the beak offered any weight advantages. Instead, he suggests that as the early birds evolved and their forelimbs became completely dedicated to being wings, they no longer had appendages for manipulating food, for turning morsels to go headfirst into the mouth. Teeth got in the way, and the wider beak, in effect enlarged reptilian scales over the top of the jaw, gave them room to slide food back and forth in the mouth before swallowing.

—JOHN NOBLE WILFORD, October 1995

Fossil "Terror Bird" Offers Clues to Evolution

THE DISCOVERY in Antarctica of the fossil beak of a giant carnivorous "terror bird," taller than a human being, is casting additional light on the role of that continent, now largely buried in ice, in the evolution and worldwide spread of species.

The bird was known to have existed in South America and as far north as Florida, but the discovery of its beak in Antarctica, first reported in 1987 and described more fully in the *Antarctic Journal,* greatly extends its range to the south.

In addition to the fearsome bird, new fossil discoveries in that frigid region include an alligator, two species of marsupial and a bird with an estimated wingspan of 17 feet, 6 feet longer than that of the Andean condor, the longest wingspan of birds living today.

Combined with new, precise determinations of when South American species reached North America, after those two continents joined, and when North American species arrived in the south, a new picture is developing of how the three separate continents each contributed to the evolution of life.

While the role of Antarctica may never be fully understood as long as most of it is ice-covered, it is evident that the continent now at the South Pole played a major role, serving perhaps as originator as well as spreader of species.

The recent fossil discoveries in Antarctica were at Seymour Island, off the tip of the Antarctic Peninsula, extending the known habitat of the giant birds and their cousins from Antarctica to Florida, although they could neither fly nor swim. The Antarctic specimens date from the period, more than 40 million years ago, when that region was warm, forested, and linked to South America and Australia.

The first indication of Antarctica's verdant past came in 1902 when Swedish geologists found fossil fig trees, laurel, beech, sequoia and evergreens that towered to 150 feet. Now Antarctica is proving a treasure trove of fossil mammals, reptiles and birds as well.

Writing in the *Journal of the Royal Society of New Zealand* in 1987, Miklos D. F. Udvardy of California State University in Sacramento, an authority on the historical relationships among species in different southern lands, said recent discoveries confirm that the Antarctic "played a central role in the evolution of all biota in the Southern Hemisphere, and indeed, in parts of the Northern Hemisphere as well."

For the period before the continents separated, Seymour Island is "surely destined" to be one of the more important fossil sites in the Southern Hemisphere, "if not in the world at large," David H. Elliot, an authority on Antarctic fossils, wrote in a new treatise on that island.

The nature of the giant carnivorous birds in Antarctica has been deduced by specialists from the remains of their cousins elsewhere. One such bird, in Florida, may have been ten to twelve feet tall. The birds were too heavy to fly and are thought to have run faster than an ostrich or horse. The bird's head was longer than that of a horse, and it presumably used its massive, hooked beak to tear apart its prey, after striking it down with one of its huge clawed feet.

The Antarctic specimen, consisting of most of a massive beak, was found by Dan S. Chaney of the Smithsonian Institution in Washington. It is described by him and by Michael O. Woodburne and Judd A. Case of the University of California at Riverside in *Antarctic Journal*. While their identification is qualified, they write that it "apparently" is of the phororhacoid group, found primarily in Argentina. It is, however, related to the cranes and rails, rather than to such flesh-eaters as hawks, eagles and owls.

Larry G. Marshall of the Institute of Human Origins in Berkeley, California, an authority on this creature, calls it the terror bird. It was, he has written, "probably the most dangerous bird ever to have existed."

He wrote recently in *American Scientist* that it was the only large carnivore in South America when a land bridge formed three million years ago, linking that continent with North America. By that time, he said, all large marsupial carnivores south of the isthmus had become extinct.

A species of the giant birds known as *Titanis* went north and colonized Florida, where its remains have been found at four sites, most recently a year ago. It has been proposed that only these giant birds were able to kill and eat the huge, armadillo-like mammals called glyptodonts, which had also migrated north at that time.

In the Ice Age Epoch, the birds became extinct, as did large numbers of other very large animals. Dr. Chaney and his colleagues estimate the age of their Antarctic specimen as 46 million years, based on the dating of seven-inch footprints on nearby King George Island. This raises the question of whether the giant birds, as well as other animal species that later migrated north, evolved in Antarctica.

So far, according to Dr. Allison V. Andors of the American Museum of Natural History in New York City, the evidence points to South America as the source, since so many varieties of the giant birds and their smaller cousins are found in deposits there. But specialists have pointed out that the oldest fossil species were already highly evolved, suggesting that they might have come from elsewhere, such as Antarctica.

While the role of that continent in spreading species to neighboring continents is only beginning to be learned, what happened when the Americas became linked is becoming much better known. "It is now possible," Dr. Marshall said, "to assess the interchange in detail, and to analyze the tempo and mode of dispersal and the rates of extinction and origination in successive faunas."

While South America was isolated, marsupials filled every ecological niche. There was a dog-like carnivore (borhyaenid) which, Dr. Chaney and his colleagues believe, may have been put out of business by the carnivorous birds. They apparently became extinct before the interchange, as did a saber-tooth marsupial (*Thylacosimilus*).

Once the bridge formed, Dr. Marshall said, the invasion was swift, many species arriving almost simultaneously. Fossils 2.5 million years old in Florida, Texas, New Mexico, Arizona and California testify to the sudden appearance of seven genera of South American land mammals. They included giant glyptodonts, two species of armadillo, two ground sloths, a porcupine and a large capybara, in addition to the giant carnivorous bird, *Titanis*.

Headed south were skunks, peccaries and horses, followed, 600,000 years later, by saber-tooth cats, dogs, elephant-like gomphotheres, bears, tapirs, deer and camels (which evolved into llamas).

Dr. Marshall believes invaders from the north were "worldly wise," because they had already sustained periodic invasions from Eurasia. The southern fauna, he said, were "predator naive," having lived in isolation for millions of years.

—WALTER SULLIVAN, January 1989

2

FAMILY LIFE
OF BIRDS

E veryone knows the edifying stories of geese that pair for life. But it turns out that birds are far from models of family life. As biologists have studied birds more closely they have found that with many species they are watching a soap opera of seduction and trickery.

It all has a serious purpose, however. The females want to get the best male genes for their offspring, the males want to insure they are raising their own chicks, not some other bird's. Out of these simple rules birds have devised a wonderful suite of bedroom farces, every bit as rich as the human variety.

Biologists Tell a Tale of Interfering In-Laws

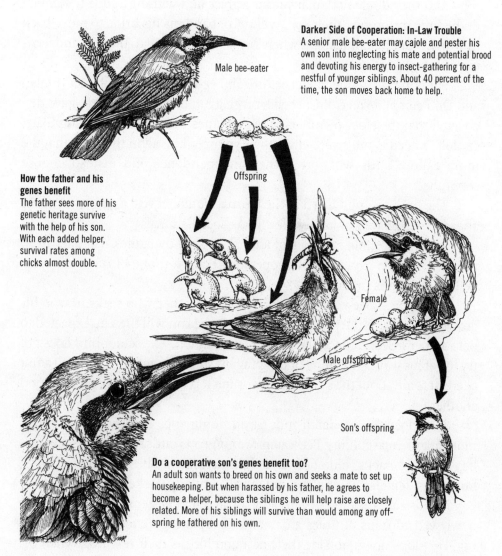

Male bee-eater

Darker Side of Cooperation: In-Law Trouble
A senior male bee-eater may cajole and pester his own son into neglecting his mate and potential brood and devoting his energy to insect-gathering for a nestful of younger siblings. About 40 percent of the time, the son moves back home to help.

How the father and his genes benefit
The father sees more of his genetic heritage survive with the help of his son. With each added helper, survival rates among chicks almost double.

Offspring

Female

Male offspring

Son's offspring

Do a cooperative son's genes benefit too?
An adult son wants to breed on his own and seeks a mate to set up housekeeping. But when harassed by his father, he agrees to become a helper, because the siblings he will help raise are closely related. More of his siblings will survive than would among any offspring he fathered on his own.

Michael Rothman

AMONG HUMANS, a visit with the in-laws can rank as one of life's little blisters, an experience just slightly more agreeable than, say, a CAT scan. But humans have nothing on the white-breasted bee-eaters of Kenya. These birds are such outrageous relations that when a young, newlywed female moves into her husband's territory, ready to lay her eggs and get a family under way, her parents-in-law will do everything in their power to wrench the young couple asunder.

And they do it with such cunning grace and winsome guile that much of the time the groom, their son, willingly abandons his bride to move back in with his mother and father, where he devotes himself to the care and feeding of his parents' new brood.

Studying two large flocks of African bee-eaters, Dr. Stephen T. Emlen and Dr. Peter H. Wrege of Cornell University in Ithaca, New York, have discovered that the elder members of the colonies will attempt to manipulate, exploit, wheedle and sweet-chirp their younger kin, all in the hope that the more callow birds will forfeit their independence and instead choose servitude.

The findings, published in the journal *Nature*, reveal a new twist in the already snarled web of family life among social animals.

These and similar studies of animal kinship dynamics also offer fresh insights into the evolution of cooperative behavior, one of the most compelling questions in biology.

In a typical bee-eater encounter, it is the father who seeks to woo his grown son back to the parental nest, where the son will then be expected to help gather insects to feed the father's latest clutch. The senior bird does not try to get his way by mauling or bullying his son or daughter-in-law; all adult bee-eaters are about the size of a thrush, and it is hard for one to push around another.

Nor does the patriarch indulge in dominance displays or flaunt his more mature masculinity. Bee-eaters are gorgeous birds, with opalescent bellies, emerald backs, blue tails and shimmering splashes of red and black; but males, females, young and old are all similarly decked out.

So instead, the father becomes a jovial but persistent pest. He visits the newlyweds dozens of times a day and disrupts their efforts at housekeeping. He plops down outside their nest and blocks their re-entry. When the son is trying to fatten up his bride in preparation for egg-laying, the father nudges in and begs for the food.

All the while, the elder bird punctuates his unseemly behavior with the many friendly little gestures of bee-eater sociability and solidarity: quivering his tail, chattering his bill, chirruping sweetly.

About 40 percent of the time the son, perhaps with a stifled sigh of resignation, concedes defeat and moves back home to help raise his siblings. The deserted female is left to languish around her own nest with little to do. She may even have already laid a few eggs, but without her mate's assistance to rear the chicks, the offspring do not survive.

The new report offers the most spectacular evidence of what Dr. Emlen calls "the darker side of cooperation," the efforts by some members of highly social animals to wrest from their relatives a degree of assistance and sacrifice extending far beyond the call of duty.

Scientists are finding that in many bird species and a few gregarious mammals like mongooses and wild dogs—societies where parents, grandparents, aunts, nieces and in-laws all breed and feed together in close quarters—some acts of what look like blissful cooperation between kin are actually subtle forms of exploitation.

The younger relative in the transaction is not always a total loser in the arrangement. In the case of the bee-eaters, the son, by helping his parents raise his brothers and sisters, keeps some of his own heritage alive indirectly, through the many genes he shares with his siblings.

Nevertheless, the younger bird would fare somewhat better from a genetic standpoint were he to raise his own chicks, and he would try to do as much if it were not for his nagging elder.

Researchers are identifying which sort of environmental and social conditions allow elder animals to manipulate the youngsters, and which conditions will encourage the subordinate creatures to rebel.

"Sometimes, when you're studying a cooperative group of kin, it all seems great and lovey-dovey on the outside," said Dr. Stuart Strahl, assistant director of Wildlife Conservation International, a division of the New York Zoological Society. "But inside it's a real social mess. And that's not surprising, is it? What do you think would happen if you moved back in with your parents?"

Biologists have long known that many species of birds and mammals engage in what is called cooperative breeding, where one lucky couple in a group freely reproduces while the other adults on the team forgo their own fecundity, instead dedicating themselves to the care and feeding of the principal pair's offspring.

That act of apparent altruism seemed to defy all evolutionary sense.

But as researchers studied these cooperative breeders, they discovered that in almost every case, the martyr adults were close kin of the breeding pair, usually children or siblings. Thus, the sacrificers were obeying at least some of the tenets of Darwinism; although they were not bearing their own babies, they were still working for the good of their bloodline.

"You are furthering the kinds of genes you tend to have," said Dr. Glen E. Woolfenden, a biologist at the University of South Florida in Tampa.

But on further scrutiny, researchers realized that the indirect explanation alone did not suffice, and that additional factors had to come into play to justify an animal's decision not to reproduce. They began to see that the nonbreeding adults often had subplots of their own going on when they opted to help around a kin's nest.

Normally, the birds who served as helpers were relatively young, and some seemed to view the season they spent working at home as a kind of apprenticeship, where they learned to rear young under the safest possible circumstances.

More often, animals became helpers when they could not find nesting areas of their own, either because surrounding territory was too crowded with competing members of their species or because most potential sites were extremely vulnerable to predators. In such cases, helpers often seemed to be playing a waiting game, assisting their elders and hoping the relatives would die off soon, leaving the breeding spot to them.

Dr. Strahl has studied another avian species that breeds in cooperative groups, the hoatzin, a bizarre, slow-moving bird found in Latin America that eats leaves, digests them laboriously with a belly like a cow's and climbs around from one plant to another. Young hoatzins are helped in their climbing by unique temporary claws that extend from the tips of their wings.

These birds have such specialized nesting and feeding needs that all the most desirable breeding habitats—leafy, well-protected little islands in swamps—are bristling with scores of hoatzins, each group a squawking, squabbling clan of relatives. Inexperienced hoatzins have a terrible time trying to break out on their own.

"Virtually all the birds have to wait two years before they can breed," Dr. Strahl said. "Often the best long-term strategy for males is to help around the house and try to inherit the territory." By the nature of the hoatzin system, the females have to disperse from their birthplace, and they spend months flying about from one territory to the next, desperately seeking an opening in a foreign clan.

Among other species, young adults may decide to help rear their kin with the hope that those newborns will then become their little acolytes when they at last are ready to propagate.

Dr. Woolfenden, who has spent 23 years observing the Florida scrub jay, has determined that birds do well to serve as helpers, particularly when most of the prime nesting spots in the neighborhood are spoken for. As the helper jays age, they try to use the kin they helped raise as soldiers to wage war against other birds and appropriate their homes.

"Older helpers in larger families have a greater probability of getting breeding territory than older helpers from smaller families," Dr. Woolfenden said. "The idea is to raise an army and then get the advantage of that army."

But in other cases, helpers obviously are on the short end in their transactions with relatives. Dr. Peter M. Waser, a biologist at Purdue University in Lafayette, Indiana, and his student Scott Creel have looked at cooperative breeding among dwarf mongooses, rat-size African mammals that are distantly related to weasels.

The creatures live in abandoned termite mounds in groups of 20 or so close relatives, with one pair of adults doing all the breeding and the rest doing everything else, guarding the den, feeding the young and carrying the little pups around.

Most remarkably of all, the subordinate females in the group also serve as wet nurses, lactating to feed the babies of the dominant female. "That's an extraordinary investment of resources," Dr. Waser said. "Milk is very expensive to produce and females will not usually produce it unless they have their own children."

The biologists have found that the dominant females wrest this dedication from their kin in one of two ways. In rare instances, young and subordinate females do become pregnant themselves, but their offspring always mysteriously disappear, very likely the victims of infanticide by the dominant couple. And once those bereaved mothers are producing milk, they might as well suckle the young that are around, greatly increasing those pups' likelihood of surviving.

But in another, more inexplicable process, subordinate females simply begin lactating just as the pups of their domineering relatives are emerging. That outpouring is a particularly impressive hormonal feat given that the helping females are often sexually repressed in other ways.

Through urine analysis of hormone levels, the Purdue scientists have found that a subordinate female has a much lower estrogen concentration when compared with the sovereign female, which is why the subordinates usually have trouble getting pregnant.

The scientists believe that the dominant female keeps the other females under constant, low-level stress, which suppresses estrogen production. Yet somehow, through a mechanism that remains to be determined, the harassed, yielding females will still generate milk.

Subordinate females might be expected to rebel against their unjust fate, except that among mongooses, patience does pay off. In general, the creatures fail miserably when they attempt to stake out their own territory. Mongooses have many enemies in the African savanna, and in fact their extreme vulnerability is probably the reason they evolved as social animals in the first place: it pays to have a network of kin around to keep watch for predators.

Over the 15 years that Dr. Waser and his colleagues have followed the mongooses, they have seen 12 cases in which pairs of young adults wandered off to establish their own homes. Of all those intrepid efforts, only one produced an offspring that managed to reach adulthood.

And beyond the safety benefits of staying at home, it seems the older the females get, the likelier they are to be given the eventual opportunity to breed. Even when the dominant female remains in the mound, still clearly in charge and still the biggest breeder of them all, she seems willing to permit the older females to have at least a few pups.

"The question is, how long can the dominant female get away with harassing the subordinates and still keep them in the group?" Dr. Waser said. "The answer seems to be that it eventually behooves her to let the subordinates breed."

Sorry as the underling mongooses' lot may be, it is nothing compared with that of a young female bee-eater whose mate has abandoned her in favor of his parents. She had had no choice but to leave her own home to seek a mate. Among bee-eaters, as with most social birds, the females disperse upon reaching sexual maturity, presumably as a built-in mechanism to keep incest rates low.

The majority do manage to begin breeding soon after leaving home, in more or less monogamous relationships. But for the female whose father-in-law manages to pull the son back under his domination, an entire breeding season is lost.

Eventually her mate comes back to her, usually by the next year, when the father is likely to have died. Yet while the son benefits indirectly by rearing his siblings, the female reaps nothing from his capitulation to his father.

"The daughter-in-law is definitely the loser in this situation," Dr. Emlen said.

She may, however, try to get her revenge. On rare occasions, a forsaken female will try to save her unborn offspring, those eggs already fertilized by her treacherous mate. She will sneak them into the well-tended nest of her parents-in-law and in a roundabout way win her husband's help after all.

—NATALIE ANGIER, April 1992

Mating for Life?
It's Not for the Birds or the Bees

AH, ROMANCE. Can any sight be as sweet as a pair of mallard ducks gliding gracefully across a pond, male by female, seemingly inseparable? Or better yet, two cygnet swans, which, as biologists have always told us, remain coupled for life, their necks and fates lovingly intertwined?

Coupled for life, with just a bit of adultery, cuckoldry and gang rape on the side.

Alas for sentiment and the greeting card industry, biologists lately have discovered that, in the animal kingdom, there is almost no such thing as monogamy. In a burst of new studies that are destroying many of the most deeply cherished notions about animal mating habits, researchers report that even among species assumed to have faithful tendencies and to need a strong pair bond to rear their young, infidelity is rampant.

Biologists long believed, for example, that up to 94 percent of bird species were monogamous, with one mother and one father sharing the burden of raising their chicks. Now, using advanced techniques to determine the paternity of offspring, biologists are finding that, on average, 30 percent or more of the baby birds in any nest were sired by someone other than the resident male. Indeed, researchers are having trouble finding bird species that are not prone to such evident philandering.

"This is an extremely hot topic," said Dr. Paul W. Sherman, a biologist at Cornell University in Ithaca, New York. "You can hardly pick up a current issue of an ornithology journal without seeing a report of another supposedly monogamous species that isn't. It's causing a revolution in bird biology."

In related studies of creatures already known to be polygamous, researchers are finding their subjects to be even more craftily faithless than previously believed. And to the astonishment, perhaps disgruntlement, of many traditional animal behaviorists, much of that debauchery is committed by females.

Tracking rabbits and ground squirrels through the fields, researchers have learned that the females of both species will copulate with numerous males in a single day, each time expelling the bulk of any partner's semen to make room for the next mating. Experts theorize that the female is storing up a variety of semen, perhaps so that different sperm will fertilize different eggs and thus ensure genetic diversity in her offspring.

Most efficiently energetic of all may be the queen bee, who on her sole outing from her hive mates with as many as 25 accommodating, but doomed drones.

Scientists also have gathered evidence of many remarkable instances of attempts by males to counteract philandering by females. Among Idaho ground squirrels, a male will stick unerringly by a female's side whenever she is fertile, sometimes chasing her down a hole and sitting on top of it to prevent her from cavorting with his competitors. Other squirrels simply use a rodent's version of a chastity belt, topping an ejaculation with a rubber-like emission that acts as a plug.

The new research, say scientists, gives the lie to the old stereotype that only males are promiscuous. "It's all baloney," said Dr. Sherman. "Both males and females seek extra-pair copulations. And what we've found lately is probably just the tip of the iceberg." Even mammals, which have never been paragons of virtue, are proving to be worse than expected, and experts are revising downward the already pathetic figure of 2 percent to 4 percent that represented, they thought, the number of faithful mammal species.

"It was believed that field mice, certain wolf-like animals and a few South American primates, like marmosets and tamarins, were monogamous," said Dr. David J. Gubernick, a psychologist at the University of Wisconsin in Madison who studies monogamy in mammals. "But new data indicate that they, too, engage in extra-pair copulations."

Scientists say their new insights into mating and the near-universality of infidelity are reshaping their ideas about animal behavior and the dynamics of different animal social systems.

"It's been a bandwagon," said Dr. Susan M. Smith, a biologist at Mount Holyoke College in South Hadley, Massachusetts. "Nobody can take monogamy for granted anymore, in any species they look at, so we're all trying to rewrite the rules we once thought applied."

Biologists say their new research suggests that many animal social systems might have developed as much to allow animals to selectively cheat as they did out of a need for animals to divide into happy couples. They propose that pair bonds among animals might be mere marriages of convenience, allowing both partners enough stability to raise their young while leaving a bit of slack for the occasional dalliance.

More than anything else, say biologists, they are increasingly impressed by the complexity of animal sexuality. "It seems that all our old assumptions are incorrect, and that there's a big difference between who's hanging out with whom and who's actually mating with whom," said Dr. Patricia Adair Gowaty, a biologist then at Clemson University in South Carolina and one of the first to question the existence of fidelity among animals. "For those of us in the field, this is a tremendously exciting time."

Researchers say that many of the misconceptions about monogamy and infidelity began in Darwin's day, when he and other naturalists made presumptions, perhaps understandable, about mating based on field observations of coupled animals. Nearly all birds form pairs during the breeding season, and biologists assumed that the pair bond was necessary for the survival of the young. Without the contributions of both males and females to feed and protect the young, experts thought, few offspring would make it to the fledgling stage. And that demand for stability, biologists assumed, likely included monogamy as well.

But as field researchers became more sophisticated in their observation techniques, they began spotting instances in which one member of a supposedly monogamous avian couple would flit off for a tête-à-tête with a paramour.

"Extra-pair copulations are called sneakers, and they really are," said Dr. Robert Montgomerie, a biologist at Queens University in Kingston, Ontario. "They're not easy to observe because the birds are very surreptitious about such behavior."

Such sightings inspired biologists to apply DNA fingerprinting and other techniques used in paternity suits to help determine the parentage of chicks. They discovered that between 10 percent and 70 percent of the offspring in a nest did not belong to the male caring for them.

Redoubling their efforts in the field, biologists began to seek explanations for the infidelities. In some cases, the female clearly was the one seeking outside liaisons.

Dr. Smith has studied the familiar black-capped chickadee of North America. She had found that, during winter, a flock of chickadees forms a dominance hierarchy in which every bird knows its position relative to its fellows, as well as the ranking of the other birds.

In the spring breeding season, says Dr. Smith, the flock breaks up into pairs, with each pair defending a territorial niche and breeding in it. Though she has determined that infidelity is rare among the chickadees, it does occur, "and in a very interesting way," she said. On occasion, a female mated to a low-ranking male will leave the nest and sneak into the territory of a higher-ranking male nearby.

"In every single case of extra-pair copulations, the female wasn't moving randomly, but very selectively," said Dr. Smith. "She was mating with a bird ranked above her own mate."

Dr. Smith suggests that the cheating chickadee may have the best of both worlds: a stable mate at home to help rear the young, along with the chance to bestow on at least one or two of her offspring the superior genes of a dominant male. "This fits into the idea that the female is actively attempting to seek the best quality genes," she said.

In similar studies, Dr. Anders Moller, a biologist at the University of Uppsala in Sweden, has found that female barn swallows likewise are very finicky about their adulterous encounters. When cheating, he said, the females invariably copulate with males endowed with slightly longer tails than those of their mates. Dr. Moller has learned that, among barn swallows, a lengthy tail appears to be evidence that the birds are resistant to parasites; this trait would be beneficial to a female's young. "Females mated to very short-tailed males engage in these extramarital affairs the most," he said. "Short-tailed males attempt to have affairs themselves, but they're rarely successful."

Some females that mate promiscuously may be gaining not so much the best genes as enough genetic diversity to ensure that at least some of their offspring thrive. Biologists studying honeybees have found that the queen bee will leave her hive only once, to mate with as many as 25 drones patrolling nearby. Tabulating her wantonness is easy: to complete intercourse, the poor drone must explode his genitals onto the queen's body, dying but leaving behind irrefutable evidence of an encounter.

And while the queen bee does have considerable reproductive demands, needing enough sperm to fertilize about four million eggs, researchers have determined that any one of the drones could provide enough sperm to accommodate her. They therefore suspect that her profligate behavior is intended to ensure genetic diversity in her brood.

But biologists say there are evolutionary counterbalances that can keep cheating in check. Females that actively seek outside affairs might risk losing the devotion of their own mates. Researchers have found that among barn swallows, a male that observes his mate copulating with other males responds by reducing his attention to her babies. Of course, males themselves are always attempting to philander, say biologists, whether or not they are paired to a steady mate at home. In an effort to spread their seed as widely as possible, some males go to exquisitely complicated lengths.

Studying the purple martin, the world's largest species of swallow, Dr. Gene S. Morton, a research zoologist at the National Zoo in Washington, has found that older males will happily betray their younger counterparts. An older martin will first establish his nest, attract a mate and then quickly reproduce, both parents again being needed for the survival of the young.

His straightforward business tended to, the older bird will start singing songs designed to lure a younger male to his neighborhood. That inexperienced yearling moves in and croons a song to attract his own mate, who is promptly ravished by the elder martin. A result is that a yearling male manages to fertilize less than 30 percent of his mate's eggs, although he is the one who ends up caring for the brood.

"The only way for the older males to get the younger females is to attract the young males first," said Dr. Morton. "The yearlings end up being cuckolded."

Older males often try to appropriate a younger male's partner. Studying mallards and related ducks, Dr. Frank McKinney, curator of ethology at the Bell Museum of Natural History at the University of Minnesota in Minneapolis, has found that males often try to force sex on females paired to other males. The females struggle mightily to avoid these copulations, he said, by flying away, diving underwater or fighting back.

"Our finding is that it's usually the older, experienced males that are successful in engaging in forced copulation," he said. "They have more skills, and capturing females is a skillful business."

Driven by evolutionary pressures, males have developed an impressive array of strategies to fend off competitors and keep their females in line as well. "In almost any animal you look at, males do things in order to be certain of paternity," said Dr. David F. Westneat, a biologist at the University of Kentucky in Lexington. Mate-guarding is one widespread strategy, he said, with the male staying beside the female during her fertile times. But other strategies result in what biologists have called "sperm wars," a battle by males to give their sperm the best chance of success.

Among many species of rodents, the last male's sperm is the sperm likeliest to inseminate the female, for reasons that remain mysterious.

Hence, several males may engage in an exhausting round robin, as each tries, repeatedly, to be the last one to copulate with the female.

In studies of the damselfly, Dr. Jonathan Waage, a biologist at Brown University in Providence, Rhode Island, has learned that the male has a scoop at the end of his genitals that can be used before copulating to deftly remove the semen of a previous mate.

In other species, natural selection seems to have favored males with the most generous ejaculation. Over evolutionary time, researchers say, this has resulted in the development of some formidable testicles.

The more likely a female is to mate with more than one male, they say, the bigger the sperm-producing organs will be.

Comparing the dimensions of testes relative to body size among several species of primates, biologists have found that gorillas have the smallest. Among the great apes, a dominant silverback male manages to control a harem of females with little interference from other males, biologists say.

Chimpanzees have the largest testes of the primates relative to body size. They are the ones that live in troupes with multiple males, multiple females, and considerable mating by all.

Human beings have mid-size testicles, further evidence, biologists say, that our species is basically monogamous, but that there are no guarantees.

But lest everybody cynically conclude that nothing and nobody can be trusted, a study has unearthed at least one example of an irrefutably monogamous animal: *Peromyscus californicus,* or the California mouse, found in the foothills of the Sierra Nevadas.

Dr. David Ribble, of the University of California at Berkeley, and Dr. Gubernick, of the University of Wisconsin, have performed extensive tests to prove the rodent's fidelity. DNA analysis has shown that the pups are fathered by a female's lifelong mate 100 percent of the time.

The scientists also have coated the female in fluorescent pigment powders to see with whom the female has contact. "The powder only shows up on her mate and offspring," said Dr. Gubernick. Mother and father split child-rearing duties 50-50, he says.

"This is an extremely unusual animal," said Dr. Gubernick. "It may be one of the only truly monogamous species in the world."

The most intrepid biologists are trying to apply the new insights about infidelity among animals to the study of humans. Some say that we are basically a monogamous species, but that the urge to cheat might have an evolutionary basis.

Babies need long-term care, which probably led to pair-bonding among humans early in our evolution, biologists say. But they suggest that a man might be driven to stray from his partner to slip a few more of his genes into the pool. For her part, a woman might philander to mate with a man who has hardier genes than those of her husband.

Dr. Robert L. Smith, of the University of Arizona in Tucson, believes that lapses in monogamy helped spawn male sexual jealousy.

"There are nasty cultural manifestations of male jealousy," he said. "Female genital disfiguration, foot-binding in China—these are mechanisms by which males have controlled female opportunities to run off and mate with other males."

But women are not entirely helpless, Dr. Smith says. He suggests that evolution has provided them with ways of avoiding being too closely monitored by men, for example, by giving no clue of when they are fertile.

"If males don't know when their mates are ovulating, they can't be so diligent about guarding their partners during that time," said Dr. Smith. "That allows women to exercise their reproductive options."

Another way that women may exercise such options, Dr. Smith suggests, is by having breasts. "In great apes, conical breasts are a signal that a female is lactating and thus has low reproductive value," he said. "By having perennially enlarged breasts, women make it ambiguous to males when they're fertile and when they're lactating," again confusing men about when to guard their partners.

—NATALIE ANGIER, August 1990

For Warblers, More Songs Lead to More Extramarital Flings

A LESSON from the birds that sing in the spring: if you wish to entice a female away from a committed relationship, you had better have a lot of tunes in your repertoire.

At least that is what seems to prompt the great reed warbler female to indulge in what ornithologists call "extra-pair fertilizations." In apparent ignorance of the biblical admonition not to "covet thy neighbor's wife," males warble away at fertile females in neighboring territories in hope of getting in on the act.

Two young Swedish scientists who have been studying this nondescript Eurasian migrant for 12 years have demonstrated that males with the most varied serenades are most likely to lure females who had already paired off with other males in nearby territories. They published their findings in the journal *Nature*.

The two scientists, Dr. Dennis Hasselquist and Dr. Staffan Bensch, wondered what besides a fling on the wing might the female get from these extra-marital matings. Certainly no help at home. Her paramours have their fun and then flit off to find other wandering females, perhaps never to be seen by her again. As is sometimes the case in human households, the cuckolded male, unaware of who sired which offspring, provides child support for all the young she produces.

Not that the cuckolded male, however socially responsible, is himself all that faithful sexually. No sooner does he copulate with the female who chooses to be his seasonal mate than off he goes to a corner of his territory and warbles away in hope of luring another and another and another, returning now and then for repeated mating with his original partner. But his fling is short-lived, for as soon as his original mate's eggs begin to hatch, he quiets down and returns to the nest to help feed the young until they fly the coop.

To determine what benefits females might derive from their extra-pair fertilizations, for seven years Dr. Hasselquist and Dr. Bensch paid daily visits to a reedy Swedish swamp during the warblers' breeding season, from early May to mid-July. They captured the birds in mist nets to band their ankles with colored rings and numbered tags and extract a bit of blood for DNA fingerprinting. By using various color combinations for the ankle bands, they soon got to know every individual in the swamp. They recorded their songs and territorial locations.

Together with their colleague at Lund University, Dr. Torbjoern von Schantz, they analyzed the DNA of the adults and offspring, which enabled them to assign parentage as precisely as any court of law would ever want in a paternity suit.

The studies showed that offspring sired by males with the largest vocal repertoires were twice as likely to survive their long journey to sub-Saharan Africa, where they wintered over, and return hale and hearty to the Swedish swamp the next spring. "It is not the number of fledglings that matters, but rather the number that come back and breed that is the evolutionary measure of success," Dr. Hasselquist said.

Dr. Hasselquist, who is now doing post-doctoral studies on redwing blackbirds at Cornell University, explained that neither the unfaithful females nor the resulting offspring seem to derive any direct benefits from the extra-pair matings. Fledglings sired by males with large vocal repertoires were no bigger than those fathered by the female's original, less sonorous mate. Nor did the more vocal males make an effort to protect mother or nest or care for the offspring. Furthermore, the females paid no more attention to the babies fathered by outsiders than to those sired by their nestmates.

Rather, the researchers concluded, the offspring sired by vocally superior males seem to be better endowed genetically. In other words, elaborate songs are the female's clue that potential seducers have genes with superior survival value. To be sure, like many a rock star, these birds do not offer much else in the way of physical attractions. Larger than an American warbler but smaller than a robin, the male great reed warbler sports "boring brown plumage that is no different from the female's," Dr. Hasselquist said.

But can he sing! Perched on top of a reed, he belts out his harsh renditions almost nonstop throughout the 20 daylight hours of Swedish springtime. The song is "spectacularly loud," easily heard about one third of a mile away, the biologist said, adding, "It's an impressive song for a small bird, if not the most beautiful."

—JANE E. BRODY, May 1996

Egret Chicks in the Nest Are Natural Born Killers

WHAT BEGINS as one of nature's most heartwarming scenes, three downy white chicks aflutter in their nest, often ends as one of its most gruesome. In a nasty case of sibling rivalry, researchers have found, the two eldest chicks in an egret nest will often attack and kill their youngest sibling, sometimes pitching it over the side. And where are mother and father during all of this? Standing idly by, preening and yawning.

Researchers say the killing of a sibling, or siblicide, as well as less extreme forms of sibling rivalry, is being recognized in increasing numbers of species, presenting the knotty evolutionary question of why so many animals regularly do in their siblings while parents do not lift a feather or paw to intervene. Now those studying the egrets, the long-legged birds often seen wading in ponds and marshes in which sibling rivalry has gone to an extreme, say they have begun finding some answers.

Dr. Douglas W. Mock, a behavioral ecologist at the University of Oklahoma in Norman, says work done by him and his colleagues suggests that the mother and father do not intervene because they are unwitting actors in a ruthless evolutionary scheme that enables them to perpetuate their genes.

In cattle egrets, there is often only enough food for two of the three hungry chicks. So the parents, the thinking goes, favor the two chicks born first, providing advantages to ensure that they will be able to fight successfully for food, dominance and safety in the nest, even if it costs their runtish third chick its life. In years of plenty, even the beleaguered youngest may be strong enough to survive to adulthood along with the rest.

After piecing together this theory during 16 years of research, the researchers say they have what may be the most convincing piece of evidence yet that these parents do indeed play favorites.

Dr. Mock and his colleagues reported recently at the Animal Behavior Society meetings in Flagstaff, Arizona, that the first two chicks to hatch count among the advantages given them by their parents a whopping dose of testosterone and other androgens in their egg yolk—as much as twice the amount found in the third chick's egg.

"We have every reason to believe that parents are collaborators in siblicide," Dr. Mock said. He noted that while biologists used the anthropomorphic shorthand of intentions for these behaviors shaped by evolution, these birds are acting on instinct rather than devising a conscious strategy.

"Now we find parents are skewing these androgens in ways consistent with collaboration with the older siblings," Dr. Mock said. "The yolk is like a bag lunch that the kid takes to school. You can either give your kid a lunch packed with steroids so he'll be a brute by recess or you can give him peanut butter."

Dr. Sarah Blaffer Hrdy, an anthropologist at the University of California at Davis who is credited with being the first to show, with her work on langur monkeys, that infanticide can be a natural, even adaptive, phenomenon, said of Dr. Mock's research, "It's extraordinary and illuminating work."

Dr. David W. Winkler, an evolutionary ecologist at Cornell University in Ithaca, New York, when told of the work, said the study would be the first account of an endowment of testosterone being passed to eggs in the wild. "It raises the possibility that parents are controlling these things ultimately," he said. "The parents may be setting all the rules of the game ahead of time."

Previous laboratory work on the domesticated canary has revealed differences in the levels of testosterone in eggs. In canaries, however, parents do the reverse, giving the most testosterone to the youngest, in what may be an attempt to smooth the playing field for struggling chicks.

When Dr. Mock, who has studied both cattle and great egrets, first brought egret eggs into the laboratory to hatch, he was dismayed to find his charges engaging in frequent, bloody battles. Sure that such brutality would not be tolerated if parents were present, he nestled the young in pillows to protect them from one another.

But when he later saw chicks in the wild, he found the same mayhem he had been trying to prevent. Chicks typically fought four or five battles a day, with prolonged fights sometimes involving over 100 thrust-and-parry exchanges.

Yet in nearly 3,000 observed battles during which parents were present, Dr. Mock said, the two did nothing to intervene. "It was deliciously counterintuitive," he said.

In the intervening years, researchers have accumulated evidence that evolution has honed a strategy in which parents appear to be maximizing the number of healthy, strong chicks they produce by putting most of their resources into the first two.

Cattle egrets begin stacking the deck by laying their three eggs one or two days apart. The first to hatch grows quickly, while the other two remain trapped in their eggs. The chick that hatches next has a similar advantage over the last one. The parents preferentially feed the dominant chicks, which are able to cow the third. The first and second hatched are the top two in the pecking order, a status that an extra boost of androgen may help them acquire. Researchers note, however, that while they suspect that added testosterone and other androgens will be linked to aggression in egret chicks, as they appear to be in canaries, exactly how these hormones affect the behavior and development of egrets remains under study.

Dr. Mock said the parents were probably "sincerely interested" in raising two chicks, given the amount of food they appear to be able to bring home. But since eggs are metabolically cheap to produce, parents lay one more egg, a strategy called "parental optimism."

The third chick then acts as a kind of insurance policy against accidents with the other two. Should it turn out to be a banner food year, the parents may be able to raise the third without too much extra effort.

Interestingly, just as egret parents, with their androgen dosing, seem to be betting on two chicks to make it, the chicks also seem to see two as the magic number. When the third chick is removed from the nest, fighting falls off and relative harmony among the remaining siblings ensues. But if researchers replace the third chick, increasing the competition for food once again, the vicious pecking resumes. Relative peace also reigns if only two hatch out.

Other birds, like osprey chicks, appear to adjust their tolerance for siblings based on their level of hunger: they fight when hungry and are peaceable when full.

Many animals, it turns out, overproduce by laying more eggs or giving birth to more offspring than they can reliably raise. Sibling rivalry and siblicide are also quite widespread, as Dr. Mock and his co-author, Dr. Geoffrey A. Parker, a theoretician at the University of Liverpool in England, write in their book, *The Evolution of Sibling Rivalry* (Oxford University Press).

Kittiwake gulls push their brothers and sisters off the steep sides of the cliffs on which they nest. Fast-developing pronghorn antelope embryos literally grow through their siblings in the womb, skewering and killing competitors with whom they would have to share their mother's nutrients. Shark embryos swim about devouring each other in utero.

Dr. Peter Nonacs, a behavioral ecologist at the University of California at Los Angeles, said that sharks, ants, bees and wasps do not just kill the siblings they are competing with for resources—they eat them, especially when food is short. "The leading idea," he said, "is that these young are like iceboxes." While dead prey cannot be stored because the carcasses would rot, a living larva—even if it is a sibling—is always fresh and ready to eat.

It can even be in your own best interests, in evolutionary terms, to be eaten. Say you are sickly or otherwise unlikely to reproduce. If by eating you, a relative carrying some of your genes will be more likely to reproduce offspring carrying some of those genes than you will if you are not eaten, you should sacrifice yourself. "But," Dr. Nonacs said, "it's unclear whether or not the ones that get eaten are doing so with smiles on their faces. In fact, I doubt it."

Even some plants engage in siblicide. The Dalbergia tree in India produces seeds in flat pods that are dispersed by the wind. The lighter the pod, the farther it can travel. The seed that develops first wreaks havoc by producing chemicals that annihilate the other seeds in the pod—its siblings—in a battle for possession of this ultralight aircraft.

Researchers note that the majority of birds and other animals do not engage in siblicide, but they say that more subtle sibling rivalry in those species—as when offspring engage in a begging competition to get food—can likewise result in the death of some siblings, though more indirectly.

Dr. Hrdy said she hoped that other anthropologists would take note of Dr. Mock's work, as humans are in some ways more like birds than they are like other primates. Most primates bear a single offspring and raise it until it is weaned, when it becomes independent. Like birds, however, humans raise several offspring of staggered sizes and ages simultaneously. But Dr. Hrdy also cautioned that extreme variability in human behavior across cultures made very simple extrapolations from bird parents to human parents impossible.

Dr. Winkler, the father of twin daughters, echoed other researchers' sentiments when he said, "These are very interesting areas to explore, in part because we feel so much in common with these birds struggling to raise their young."

Dr. Scott Forbes, an evolutionary ecologist at the University of Winnipeg in Canada, said he saw evidence already for intense sibling rivalries in human families that are similar to those seen in other animals. In what is known as the vanishing twin syndrome, a large fraction of pregnancies that start with twin embryos end up with only a single baby at birth. "This is terribly suspicious," Dr. Forbes said.

Long gone are the days when researchers could view a family and see a simple, harmonious constellation of parents and babes, or even when researchers believed that pulling a single thread might unravel the tangled family knot. They say sibling rivalry is, at best, only one aspect of the family dynamics in which each player has different evolutionary alliances and conflicts with its family members. Researchers are now studying whether egret chicks are all full siblings or if they can instead have different fathers, a complication that would add an entirely new layer to this story.

"No one explanation's going to do it all," Dr. Winkler said, "and it's not going to be simple."

Researchers, meanwhile, ponder these killer chicks and their parents and try to weigh their myriad evolutionary options, which include whether to kill or not to kill.

"A sibling is a glass half full and a glass half empty," Dr. Mock said. "The study of sibling rivalry is a great crucible in which to study the limits of selfishness."

—CAROL KAESUK YOON, August 1996

Designing Birds Impress Their Mates with Fancy Decor

COME INTO MY PARLOR, beckons the male bowerbird to females that stop to appraise the mating courtyard he has constructed. See my fine bower, he suggests, a perfect avenue and platform of sticks, and see all the decorations I've gathered: bright blue feathers so rare, blue straws and bottle caps and clothespins I stole from someone's line, along with lovely leaves and snail shells, all carefully arranged on my parlor floor.

And, he might add, aside from humans, no other creature in the animal world rivals my ability to assemble such a structure simply to please my mates. In the bird world in particular, nearly all males rely on inherent qualities like coloration, song, feathery displays and aerial antics to woo females, although a few offer fruits, feathers or flowers to prospective mates.

The architectural and decorative skills of bowerbirds have intrigued naturalists since Charles Darwin's day. The structures they build are striking edifices, some as high as nine feet, and decorated with collections of natural objects arranged in distinctive patterns and colors. Indeed, some 19th-century explorers mistook the bowers as man-made, and many others assumed they were merely nests.

"The bowerbird fascinates people because it exhibits behaviors reminiscent of what we do to advertise ourselves and attract mates," said Melinda Pruett-Jones, a biologist who has spent thousands of hours studying bowerbirds. "The bowerbird can extend what nature gave it by using secondary sex characteristics the way young men wear special clothing and drive fancy cars to attract girls."

To learn more about these "Picassos of the bird world," as Dr. Jared Diamond of the University of California describes them, a handful of researchers have sacrificed physical comfort and sometimes risked life and

Satin bowerbird
The male bird first erects two walls of sticks, then lays down a layer of yellow twigs on which he places piles of orange flowers, blue objects like parrot feathers and small yellow leaves. He arranges small snail shells near the wall of sticks and large shells at the periphery. The avenue is oriented north to south.

Michael Rothman

Golden bowerbird
The male bird finds a natural vine that works as a crosspiece between two tall adjacent saplings, using the crossbar as a display perch. He constructs the bower with sticks, sometimes piling them nine feet high. Yellow flowers and lichens provide additional decoration.

Striped gardener bowerbird
The bower is a hut built with sticks woven around a central maypole. The male uses brown leaves around the base and groups decorative objects by color: blue beetle wings, yellow flowers, orange flowers.

Michael Rothman

limb, trekking through dense mountainous jungles in a remote part of the world once populated by headhunters.

After decades of casual observations and speculation, recent scientifically designed studies are yielding a deeper understanding of bowers, especially their evolutionary significance and the role they play in sexual selection. At the same time, the bowerbird has become an important case study in theories of how females choose males.

"Currently, there is a resurgence of interest in the function of bowers," said Dr. Diamond, a physiologist at the University of California Medical School in Los Angeles, who has trudged into the jungles of the Indonesian part of New Guinea year after year to explore the biological meaning of these edifices.

For example, Dr. Gerald Borgia, an evolutionary biologist at the University of Maryland, has enlisted 75 volunteer assistants to monitor activity in bowers. They have observed thousands of bowerbird courtships and copulations.

To consummate the relationship, the male must move behind the female. Dr. Borgia found that the walls of bowers actually reduce the male's ability to capture uninterested females since they offer the females a means to escape should the male try to jump them. Why then would the male bother to build these walls of sticks? he wondered.

"The stick bower seems to increase recruitment of females, who are more willing come to visit the bower, see the male's show and decide if they want to mate," he said in an interview. "So while the male may get to mate with a smaller percentage of females who come to check him out, many more females visit the bower and the total number of matings is higher.

"It's like a loss leader in the grocery store," Dr. Borgia said. "It gets the traffic in and you end up doing better business."

To spy on bower activity, Dr. Borgia and his assistants set up cameras at 30 bowers, with the cameras designed to roll whenever a bird crossed an infrared beam.

Until studies like Dr. Borgia's, much that was known about bowerbirds was anecdotal, and a lot of those "facts" turned out to be wrong.

Of the 18 species of bowerbirds, all denizens of Australia and New Guinea, 15 are known to build bowers. There are several types. The satin, Australian regent and great bowerbirds build a simple avenue as their bower. The yellow-breasted bowerbird consructs a double avenue. Macgregor's bowerbird of New Guinea weaves a maypole-style bower of sticks around a sapling or fern and surrounds it with a raised circular courtyard.

Some bowerbirds elaborate on the maypole style by constructing huts nearly five feet high around it. And the golden bowerbird decorates adjacent saplings with sticks, sometimes piling them to a height of nine feet, and uses a crossbar between them as a display perch.

The least elaborate structures are the wall-less mat bowers of the tooth-billed bowerbird. This bird decorates a clearing with leaves more than a foot long, all carefully turned upside down to create a contrasting stage for the male's brief courtship dance.

Dr. Borgia, who spent three years studying tooth-billed bowerbirds, was the first to observe their copulatory behavior, which is tantamount to rape. When females come into the bower, the male doesn't bother with courtship: he simply jumps them. Cameras positioned this year by Dr. Borgia at 21 bowers showed that about 70 percent of female visits resulted in copulation.

Most other bowerbirds, he noted, "are much more polite; they wait for the female to show she's interested." Less than 8 percent of such visitations result in mating. Yet many more females visit the avenue-style bower, which means that the courteous architects of these edifices do not lose out in the mating game.

Another mat-builder is Archbold's bowerbird, which constructs a courtyard of lichens and decorates it with piles of snail shells. However, Archbold's, a little known New Guinean species, has an added attraction. Dr. Dawn Frith and Dr. Clifford Frith, two Australian naturalists who discovered the Archbold's bowers, found them decorated with "borrowed" head plumes from the male King of Saxony bird of paradise. The Friths note that this is no small feat: the bird of paradise is rare, lives in nearly impenetrable jungle and molts its two magnificent head feathers only once a year.

Archbold's bowerbirds also drape trails of orchids over their bowers, daily replacing fading flowers with fresh ones. Some bowerbirds gather huge leaves several times their length. Others adorn their bowers with lichens and mosses; butterfly wings and beetle skeletons; brightly colored fruits and stark black fungi; red, white or blue flowers; berries; insect frass; seed husks; and yellow leaves.

And those that dwell near human habitats selectively scavenge artifacts, from ballpoint pen tops and "lost" bracelets to toothbrushes and red shotgun cartridges, that match the bird's color scheme. One particularly brazen individual even untied and struggled to remove a researcher's blue shoelaces as loot for its bower.

To the outside observer, "the bower decorations often look like a pile of junk," said Glen Threlfo, an Australian nature photographer who has chronicled the life cycle of Australian bowerbirds from courtship through nestlings. But naturalists correctly inferred that the bower trinkets must mean something to potential mates. Along with the bower structure itself, they presumably convey to females that the owner is a desirable male.

More than 30 years ago, E. Thomas Gilliard of the American Museum of Natural History noted that the less colorful the male bowerbird, the more fanciful his bower. Bowerbirds seemed to have evolved a system of sexual selection that transferred the female's attention from personal appeal of the male himself to the attractiveness of his proverbial etchings.

But careful studies made recently by American researchers indicate artistic talent alone is hardly enough to seduce the female bowerbird. Rather, it is the finesse of the bower's construction, its state of repair and the desirability, rarity and quantity of its accouterments that together inform the discriminating female whether she should let herself be beguiled by its owner.

As Dr. Borgia pointed out, the female bowerbird gets nothing from the polygamous male except his genes. He does not protect the nest, procure food for the young or even visit his family after mating. Thus the female's goal, in Dr. Borgia's view, must be to ensure that her chosen Don Juan will bestow upon her offspring genes with top-notch survival value. In other words, females would naturally be drawn to dominant males.

Mrs. Pruett-Jones of Chicago's Field Museum and her husband, Dr. Stephen Pruett-Jones, an evolutionist at the University of Chicago, discovered wide variations in the bowers of a single species. Since 1980, they have spent thousands of hours in eastern Papua New Guinea observing Macgregor's bowerbird, an olive-brown bird the size of a robin. The maypole-building male has a bright orange crest to enhance his allure.

"No two bowers are alike," Mrs. Pruett-Jones said in an interview. The maypole can range from one to 10 feet in height, and the decorations, which vary widely, may include bits of charcoal, fungi, seeds, fruit, insect frass, pandanus leaves, lichens and the parts of iridescent insects.

"We have counted as many as 500 decorations at a single bower," they reported in *Natural History* magazine. Dr. Diamond has noted that the combined weight of the towers and ornaments of the Vogelkop or gardener bowerbird can reach 70 pounds.

Researchers have found that birds of various ages build bowers, but those of young males are crude "practice" bowers. Only fully mature males among these long-lived birds build bowers likely to attract females. Young males seem to learn from their elders by observing the construction techniques of master builders.

The effort to enhance his bower's attractiveness and to diminish competition for mates keeps the male bowerbird fully occupied. He must divide his time between trying to loot and destroy the bowers of nearby males and protecting his own bower from the competition.

Whenever a bower is attacked, its owner must scurry to repair it and replace any stolen decorations. The more time he must spend repairing his bower, the less time is available for courtship and mating.

Given the rigors of this competition, it is no surprise that a well-constructed and heavily decorated bower sends a powerful message to females that the owner is a mature, clever, dominant, aggressive male that is likely to contribute genes with high survival value.

The bower of the gardener bowerbird, Dr. Diamond said, is the most elaborate structure erected by any animal other than humans. To construct a good bower, he wrote in *The American Naturalist,* "a male must be endowed with physical strength, dexterity and endurance, plus searching skills and memory—as if women were to choose husbands on the basis of a triathlon contest extended to include a chess game and sewing exercise."

Not all bowerbirds rely on thievery and bower attacks to prove their mettle. Dr. Borgia has found that spotted bowerbirds space their avenue bowers too far apart for there to be much interaction between neighboring males. Instead of wooing females with the quality of bowers, he said, "the male puts on a very energetic display—running three meters from the bower, charging back in, throwing decorations and using its aggressive call for courting."

Since other bowerbird species that engage in stealing and bower destruction have much more sedate displays, Dr. Borgia concluded that "the athleticism of the spotted bowerbird compensates for the fact that his bower doesn't reveal much about his quality to potential mates."

According to several studies, the birds experiment, like professional decorators, with the choice and placement of bower art. Dr. Diamond offered bowerbirds poker chips of seven different colors. He found that the birds have definite color preferences. They moved the chips around, placing certain colors inside their bowers and others in the courtyard outside. Some threw all the colors out.

Three bowerbirds are known to be bowerless. Unlike the polygamous bower builders that assume no parental responsibilities, the bowerless species help defend the nest and find food for the nestlings.

Other species thought to be bowerless turn out to be adroit concealers. The fire-maned bowerbird hides its bowers in a remote jungle high in the Adelbert Mountains in Papua New Guinea, according to Roy Mackay, an Australian naturalist whose studies have been supported by the New York Zoological Society.

Mr. Mackay said his quest for the bower took four visits to the Adelberts, during which he had to endure false leads from the local villagers, hundreds of mosquito bites, unseasonably heavy rains and devastating attacks of dysentery. Eventually a woman who had stumbled upon a bower while hunting the previous week led him to it. He found an avenue bower simply decorated with seven Prussian-blue berries and one large blue and white berry. A few days later, he said, a spotted catbird dove in and swallowed all the berries.

On a later visit he finally saw the bird in full display. It fanned its fiery nape and mantle feathers into a sort of platform above its eyes. Then it hopped near the female trying to impress her with its finery. After many hours of patient endurance in the steamy jungle, it was for Mr. Mackay an ample reward.

—JANE E. BRODY, March 1991

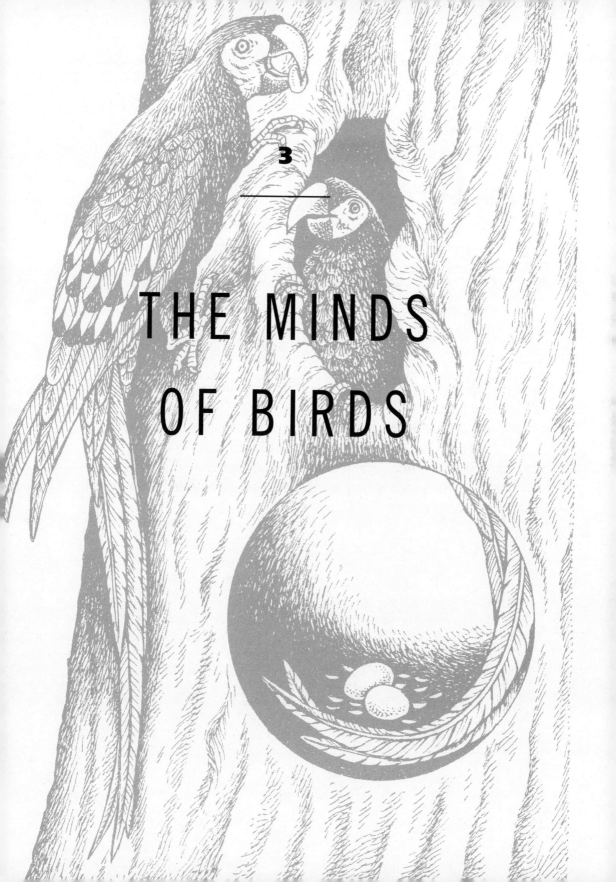

THE MINDS
OF BIRDS

3

Birdbrained, not a term of praise, is a phrase no doubt inspired by the familiar behavior of chickens and turkeys. But these domesticated birds have had many aspects of their wild nature bred out of them, and are hardly a fair representative of avian cognitive abilities.

In fact, as the articles in this section relate, biologists are gaining increasing respect for the mental capabilities of birds. Crows are a particularly intelligent family. Recently a species of crow has been found to use tools at a level of sophistication at least equal to that of chimpanzees. A remarkable gray parrot named Alex appears to have unusual intellectual abilities, at least in the hands of its trainer.

Researchers interested in how the brain works have turned to birds because of the ability of certain species to learn songs or remember the location of winter food caches.

Birds clearly have minds with many interesting features and capabilities.

Second Greatest Toolmaker? A Title Crows Can Crow About

THE COGNITIVE ABILITY to design, make, standardize and use tools is widely thought to be a hallmark of human society, exceeding the capacity even of chimpanzees, mankind's brightest primate relatives. But a biologist who has spent three years studying a breed of crows in South Pacific rain forests reports that the birds actually make tool kits to extract worms and other prey from holes in trees and dead wood.

The toolmaking ability of these crows, he believes, is superior to any observed in other nonhuman species.

All corvids, members of the crow genus, exhibit innate ability to solve many kinds of problems. But according to Gavin R. Hunt, a biologist at Massey University in Palmerston, New Zealand, one species is special: *Corvus moneduloides* of the New Caledonia island group 900 miles east of Australia.

In a paper in the journal *Nature,* Mr. Hunt said he had observed that "crow tool manufacture had three features new to tool use in free-living non-humans, and that only appeared in early human tool-using cultures after the Lower Paleolithic: a high degree of standardization, distinctly discrete tool types with definite imposition of form in tool shaping and the use of hooks."

Claims by scientists to have detected highly intelligent behavior in animals are often challenged by skeptics, and Mr. Hunt said in an interview that he expected sharp questions from his peers.

In a comment published in the same issue of *Nature,* Dr. Christophe Boesch, a zoologist at the University of Basel, Switzerland, questioned whether the tools Mr. Hunt observed crows making and using are truly planned or are merely shaped by trial and error for specific tasks. Only if the shape of the tool is preconceived by its makers can the process be con-

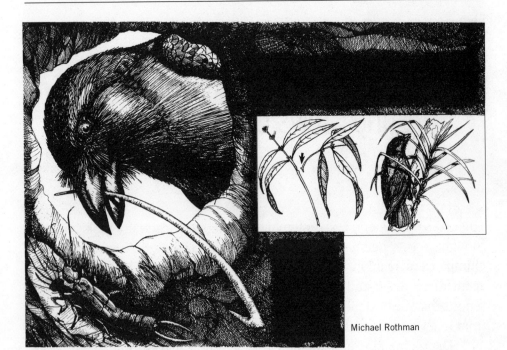

Michael Rothman

Custom Tools Made for Special Tasks
A researcher says that the crow *Corvus moneduloides* makes hooks and toothed probes to help it in its search for hidden insects. The hook is made by nipping a twig at the base and then removing leaves and bark to leave a smooth tool. The toothed probe is snipped out of a pandanus leaf; it is pointed and goes in a hole, and when the teeth come out they drag insects along with them. Above, the crow uses its tool to seek out an earwig lurking behind bark.

sidered "by some to be the characteristic of the existence of culture," Dr. Boesch wrote.

But whether crow toolmaking is planned or not, he added, Mr. Hunt's "fascinating paper gives much food for thought and argument," showing at least that "tool use in birds is less stereotyped than previously thought."

A more positive assessment came from Dr. Randall L. Susman, an anatomist at the State University of New York at Stony Brook, who has extensively studied the anatomy and behavior of wild chimpanzees and other intelligent primates in the African wilderness.

"I'm not a psychologist," he said, "but if the birds Mr. Hunt has observed are actually shaping implements according to some plan, I'd call their activity cognitive. The only higher primates that make tools conforming to a preset template are human beings." Although chimpanzees use objects they find as tools, if they modify the objects at all, it is not according to any standardized pattern, he said.

During his research from 1992 to 1995 in four mountain rain forests of New Caledonia, Mr. Hunt watched *moneduloides* crows make and use two distinctly different types of tools, one of them hooked at one end, and the other made from a tapered piece of stiff leaf from the pandanus plant, with a barbed edge on one side.

To make hooked-twig tools, he said, the crows use their wide beaks to carefully pull a twig away from a branch using a "nipping cut" to create a distinct hook at the twig's end—the end the bird inserts into holes. Holding the twigs with their claws and shaping them with their beaks, the crows remove leaves, carefully shape the hooks and trim off the bark to make their tools smooth.

A second type of tool manufactured by the crows uses pieces cut from the stiff, jagged-edged leaves of pandanus plants. In finished form, these tools resemble locksmiths' picks, tapered to points and with serrated barbs along one edge; Mr. Hunt calls them "stepped-cut tools." To make one, a crow takes successively deeper bites from the section of leaf while it is still attached to the plant, and then bites off the finished implement. When the pointed end is inserted into a hole, the natural barbs along the edge of the leaf point outward so that withdrawing the tool snags and pulls up the prey.

Moneduloides are small crows resembling European jackdaws. They have broad bills with which they grasp their two types of tools in different ways. The hooked twig is held at an angle to the bill and the crow moves its head from side to side to probe a hole. To use a stepped-cut tool, the crow holds it by the broad end with the tip pointing straight ahead. The bird probes holes with it by moving its head back and forth.

Mr. Hunt compared the lengths and number of stepped cuts in tools made by *moneduloides* crows in three different areas, and found significant differences among them, perhaps suggesting cultural differences among neighboring crow communities analogous to differences among early human societies in the ways they shaped stone spear points.

Moneduloides crows evidently value their tools and try to keep track of them, Mr. Hunt said. When crows change their foraging sites they generally take their tools with them, he said, and when crows eat they generally grasp their tools in their feet. Sometimes crows leave their tools on secure perches while searching distant hunting grounds, returning later to retrieve their hooked twigs or stepped-cut leaves.

One of the few animal users of tools is the woodpecker finch, or *Camarhynchus pallidus,* one of 14 distinctive species of Darwin finches that evolved in the Galapagos Islands and are named for their discoverer, Charles Darwin. But the woodpecker finch does not make its tools. It plucks needle-like thorns from local plants and without modifying them, it uses the thorns as picks to tease prey out of holes. The brown-headed nuthatch of the Eastern United States uses bits of bark to probe holes for food, Mr. Hunt said.

Crows and sea gulls sometimes open whelks and other shellfish by dropping them from a height onto rocks, and sea otters use stones as tools to pound the shells of crabs and other prey held against their chests. But Mr. Hunt and many other scientists regard this as different from making tools.

In the wild, pygmy chimpanzees in East Africa modify stems or twigs to "fish" for termites in holes, Dr. Susman said, and in the Tai Forest of Côte d'Ivoire in West Africa during periods of drought and food shortage, he said, chimpanzees use stones to pound open hard nuts that are not normal components of their diet. But the shaping of stone nutcrackers by continued use appears to be accidental, with no plan in mind for making tools according to a standardized pattern, he said.

Nevertheless, the use of stones for cracking nuts seems to vary with different chimpanzee societies; on one side of the Sassandra River in Côte d'Ivoire, Dr. Susman said, the chimpanzees crack nuts with stones, and on the other side, they do not.

Chimpanzees in the wild modify leaves somewhat haphazardly to use as sponges and as umbrellas. Although this activity does not meet all the criteria for planned tool manufacture, Dr. Susman said, "there no longer seems to be a sharp dividing line between the cognitive behavior of apes and human beings." It may be, he said, that the distinctions between bird and human planned behavior has also been blurred; in any case, "humans are part of the continuum of animal cognition, not separate from it."

If New Caledonian crows intentionally modify their tools to specific shapes before using them, he said, their behavior is "qualitatively different from chimpanzees, and I'd call it cognitive."

Zoologists are revising traditional views of the relative importance of genes and learned behavior in young birds.

Experiments have shown, for example, that birds are born with an innate ability to sing the songs of their species, but that young birds need the examples provided by their elders to master the fine points of avian melody; a laboratory bird raised in isolation from others of its species never becomes expert in the songs of its race.

Mr. Hunt said that he had not noticed any evidence that *moneduloides* chicks learn the art of toolmaking from more experienced birds, but he was not able to investigate the question.

"One thing this kind of study does," he said, "is to impart a feeling of humility and greater respect for animal cognition."

—MALCOLM W. BROWNE, January 1996

Michael Rothman

When crows change their foraging sites they generally take their tools with them.

To Remember Seed Caches, Bird Grows New Brain Cells

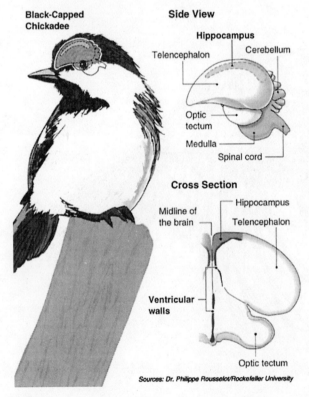

Chickadee's Brain-Renewal Project

Chickadee's hippocampus grows many new neurons each year at seed-hiding time. New growth forms along ventricle walls (below right).

Black-Capped Chickadee

Side View

Hippocampus

Telencephalon

Cerebellum

Optic tectum

Medulla

Spinal cord

Cross Section

Midline of the brain

Hippocampus

Telencephalon

Ventricular walls

Optic tectum

Sources: Dr. Philippe Rousselot/Rockefeller University

Jody Emery

EVERY FALL, when the trees of North America's forests shake themselves into bare brown bones, and the sky grows gray and surly, and the meadows offer little but brambles and shriveled berries, the little black-capped chickadee rises brilliantly to the challenges of a callous world.

The insects it has feasted on throughout the spring and summer are all dead, which means the bird must start foraging over a much wider terrain for fast-vanishing seeds and nuts. And it must join forces with huge flocks of other hungry chickadees, far more birds than it had to tolerate during the balmy breeding season.

The creature must also store food for the dead of winter, distributing little stashes in many hiding places to ensure that it does not lose everything in one big theft; and it must keep in mind where all those seeds are cached.

Not to worry: each year, when the need to sharpen its thinking is greatest, the black-capped chickadee grows a fresh new brain.

Reporting in *The Proceedings of the National Academy of Sciences,* Dr. Fernando Nottebohm of Rockefeller University in New York and his colleague Dr. Anat Barnea, now of Tel Aviv University in Israel, announced that in black-capped chickadees the hippocampus of the brain—a region thought to be critical to memory storage and spatial learning—swells with a burst of neuronal growth each October. The scientists have found that just at the time when the birds are confronting profound changes in their landscape, their social milieu and their memory requirements, they have a huge turnover in hippocampal neurons, the old cells dying off and new ones taking their place.

By comparison, chickadees kept in comfortable captivity and provided with all the seeds they can swallow display only half as much neuronal turnover as their free-foraging peers. The discrepancy suggests that the need to get very smart in a hurry stimulates the birth of new brain tissue.

The latest results add an important new dimension to the heatedly debated field of neurogenesis, the notion that the adult brain, far from being incapable of cell growth or self-repair as scientists long believed, can under certain circumstances shuck off the old and fashion the new.

In previous experiments with songbirds, Dr. Nottebohm and his co-workers demonstrated that parts of the brain believed responsible for song learning undergo annual cycles of cell death and regrowth, very likely as a way of allowing birds like canaries to forget last year's tunes and instead master new ones. Now, the scientists have shown that neurogenesis also takes place in the songbird's hippocampus, presumably as a way of keeping the creature's foraging skills and territorial maps up to date.

"This adds strength to our original hypothesis that there is a correlation between neuronal turnover and learning," he said.

Neurogenesis has also been observed in mammals, particularly rodents, although the reasons for the turnover of brain tissue in these animals remain unclear. Nor does anybody know whether neurogenesis ever takes place in adult human brains, though they suspect the answer is usually, alas, no.

Researchers hope that if they can unravel the details of neuronal regeneration in birds and other animals, they may figure out a way of coaxing forth the trick in patients suffering from neurodegenerative disorders like Alzheimer's disease, Parkinson's disease, paralysis and others.

Beyond any long-term clinical relevance, the latest experiments also raise the provocative question of how one defines a unit of memory—that is, what, exactly, is a memory made of? Most neurobiologists suspect that the unit of memory is the synapse, a connection between one brain cell and the next. They propose that the brain learns a new task or forms a memory by strengthening synaptic tendrils between neurons, or by adding new circuits to the network.

Based on his work with songbirds, Dr. Nottebohm believes otherwise. The fact that songbirds display regular cycles of neuron death and replacement suggests to him that synaptic connections are not enough to make serious, long-term memories. "If the synapse were the unit of memory," he said, "then the animal wouldn't need to sacrifice all these neurons" to build its annual mental portrait of the autumnal world. It would simply bump up the number of synapses to accommodate incoming information.

Dr. Nottebohm theorizes instead that the unit of memory is the entire neuron, and that a memory becomes stable and fixed only when an immature brain cell takes the leap toward final, and irreversible, differentiation. In assuming its ultimate fate, the neuron learns what it must learn, and keeps that memory until it dies. "The need to replace the whole neuron to make room for new memories is what's driving the system" in songbirds, he said.

Birds more than any other fauna require neuronal recycling, he said. Black-capped chickadees, for example, are about the size of mice, but they live far longer, about 12 years or so compared with the rodents' 18 months. That means birds require enough brain power to get them through a dozen difficult seasons. However, their airborne ways limit the potential size of their brain. They cannot meet the demands of their long lives by entering the world with weighty minds and copious room for new knowledge. So, instead, he said, they have evolved the mechanism of a featherweight brain that turns over easily, discarding what no longer applies and bringing in a fresh batch of impressionable brain cells.

Despite the unusual neural needs of birds, said Dr. Nottebohm, the fundamental premise could apply to all brains: to create good, strong memories that will last for months or years, they must devote whole neurons to the task, not merely stitch together a few synapses. "I think we're staring at a very basic principle here," he said. "The avian system may be a superb system to identify factors that are important for learning."

The implications of Dr. Nottebohm's ideas for humans are not entirely appealing. If he is right that the unit of memory is the entire neuron, rather than the synapse, then the aging brain with its neurons largely differentiated and locked into place looks considerably less flexible than the youthful brain, with its many unschooled neurons available to master novel skills. Humans can sprout new synaptic connections throughout life, but unlike chickadees they must largely make do with the neurons they were born with. If synaptic growth is in fact an inferior way to form memories and learn new tasks, then all those chirpy notions now in vogue that one can stay mentally fit by reading books, taking adult-education classes and keeping those dendritic trees ever-branching outward may be so much wishful thinking.

Instead, one's mental options and intellectual suppleness may start to shrink later in life as surely as does one's height.

"As you age, you probably have to manage to learn stuff in terms of the things you already know," Dr. Nottebohm said. "If you're good at metaphorical thinking, you can get away with people believing your mind is still young.

"But this may be why young people are so impatient with the wisdom of their elders," he added. "That wisdom has the flavor of things being interpreted with the ways of another era, which no longer seem relevant today."

Other neurobiologists praise Dr. Nottebohm for his elegant and often profound experiments, but they question how applicable his results are to species other than birds. They also insist that he must offer stronger proof that the turnover of neurons in a bird's brain is directly related to the animal's acquisition of a new skill or memory, rather than serving some other purpose, or no purpose at all.

"I've told Fernando that what he's doing is wonderful poetry," said Dr. Eric R. Kandel, a neurobiologist at the College of Physicians and Surgeons of Columbia University in New York. "There are lots of people who find that the world works exactly as everybody thinks it does, and then there are the very few who try to think about it in a completely different way. Fernando is one of the few."

Nevertheless, Dr. Kandel said: "I'm skeptical that cell turnover is used as a mechanism for memory, though I'm keeping an open mind. The ball is in his court now to prove it to the scientific community."

But neurobiologists confess that the scientific community's grasp of how learning and memory occur in any species is primitive and incomplete. For example, one of the favorite buzz phrases in neurobiology today is long-term potentiation: the idea that memories are laid down in the brain when a particular synaptic circuit is used enough times to become sensitized, and thus is far more likely to fire again in the future under the slightest stimulus. But trendy though the topic is, said Dr. Fred H. Gage, a neurobiologist at the University of California at San Diego, "there's very little evidence supporting the proposal that long-term potentiation is the correlate of learning and memory."

Apart from their poetic implications, the new experiments on birds are unusual for their reliance on free-ranging chickadees confronting the vagaries of nature, rather than lab-reared specimens incapable of finding seeds not presented to them on plastic dishes. The work was done at the Rockefeller University Field Research Center in Millbrook, New York, 1,200 acres of meadows, woods and marshes 80 miles north of Manhattan, where chickadees live in abundance. Studying the birds' habits, the scientists learned that in the fall the chickadees expand their home range from 3 to 30 acres and likewise expand their acquaintanceship with other chickadees, and that they begin storing their food.

For two years, the scientists captured 74 adult chickadees of both sexes and injected them with a radioactive tracer compound called thymidine. The compound is designed to be incorporated into any cells that divide, including brain cells. The researchers set the birds free again and recaptured 42 of them six or more weeks later.

For comparison, a group of chickadees living in an outdoor aviary and provisioned with food were also injected with the radioactive thymidine.

All the birds were eventually killed, their brains sliced into very thin sections and the sections scrutinized for evidence of thymidine, the signature of cell division. The scientists discovered that the greatest amount of new neuronal growth occurred in one region of the hippocampus, along the so-called ventricular wall. Because the size of the hippocampus had not increased, they assumed that old cells must have died off to make room.

They determined that cell turnover in the hippocampus continued throughout the year, but that it spiked most sharply in October, just as the conditions in the field were shifting most radically. At that point, almost 2 percent of the neurons were being replaced each day, an astonishingly high volume.

Among the aviary chickadees, the October hippocampal peak was significantly smaller.

Dr. Nottebohm points out that hippocampal damage has been linked to memory impairment in humans, monkeys and rodents, and he believes that the brain locus very likely plays a role in memory in chickadees. The temporal correlation between changes in the bird's hippocampus and changes in the world around it are too great, he said, to be coincidental. He next proposes comparing chickadees with other birds that do not store seeds, or with those who live in evergreen forests that do not change their appearance greatly each fall, to see if brain growth is meatiest in those who are neediest.

—NATALIE ANGIER, November 1994

Brainy Parrots Dazzle Scientists, but Threat of Extinction Looms

Patricia J. Wynne

Deforestation is threatening macaws who need tree holes for nesting.

IN ANCIENT INDIA, where they appeared in the earliest literature, they were cherished as symbols of love: the *Kama Sutra* stipulated parrot training as one of 64 practices men had to master. The Romans taught parrots to speak, festooned them with ivory, kept them in golden cages and sometimes valued them more highly than slaves.

No other bird seems more human, more intelligent or more affectionate. Take a baby parrot out of the nest and it will not only learn human speech but also, given a chance, become so attached to its keeper as to become a second shadow. Par-

70

rots can also be willful and demanding, jealous and contrary. They may screech and yell at anyone who walks into the room, and they can imitate any sound—a dripping faucet or a ringing telephone or the barking of a dog.

And yet for all the rich lore on parrots, biologists are only now beginning to penetrate their unexplored world, both in the laboratory and in the wild. They are finding that parrots are even more intelligent than believed, with mental abilities that may equal those of chimpanzees and dolphins. Researchers are learning that they can deal with abstract concepts, communicate with people, understand questions and make reasoned replies rather than merely, well, parrot human speech.

But even as some of the parrot's wonders and secrets are yielding to new scrutiny, nearly a quarter of the world's 300 parrot species—and nearly a third of Western Hemisphere species—are at risk of extinction.

For some species, including the splendid macaws, the reproductive rate is so low that they are poorly equipped to face what may be the most serious threat ever to their existence.

This is partly because humans are destroying the tropical forests where parrots live. The birds mostly dwell in holes they find in trees, and the destruction of the forest increases the already high odds against reproductive success for some species.

Parrots are also victims of the very fascination they hold for humans. In greater demand than ever as pets, they are being trapped by the millions. Many of the birds die in transit. The United States alone imports at least 250,000 parrots a year, according to Traffic U.S.A., the trade-monitoring arm of the World Wildlife Fund in this country, which says that the vast majority of imported parrots come from the wild. The big macaws of South America and the impressive crested cockatoos are particularly prized as status symbols. And the rarer a species gets, the more valuable it becomes and the more avidly it is sought.

A number of American conservation groups and representatives of the captive breeding industry, led by the World Wildlife Fund, have proposed national legislation to phase out the importation of exotic wild birds over five years. "Of paramount concern to us are parrots," said Ginette Hemley, the director of Traffic U.S.A.

There are cockatoos and cockatiels, parrotlets and parakeets, lorys and lorikeets, amazons and macaws. There are caiques and conures, keas and kakapos, lovebirds, budgerigars and just plain parrots. Whatever the variety, it is especially tempting, despite the danger of being misled, to assign human characteristics to parrots.

Many species live about as long as humans. A middle-age person buying a large parrot as a pet is unlikely to outlive it.

Most parrots mate for life, although divorce is common in some parrot societies. Mates go everywhere together and—unlike most other birds, which avoid physical contact except in coition—parrots spend much of their time snuggling up and preening each other. They also like to have their heads tickled.

These are among many pieces of lore contained in a book that is part of a general surge of interest in parrots. The book, widely praised by experts as an accurate survey, is *Parrots: A Natural History* (Facts on File, 1990), by Dr. John Sparks, a zoologist who heads the natural history unit of the BBC, and Tony Soper, a writer and filmmaker with the unit.

Parrot parents invest a lot of time raising their offspring. Like monkeys and apes, young parrots undergo training in family groups, where the wisdom of the elders is transmitted. In some species, fledglings join nursery groups. Like primates, parrots play, some scientists believe. Why else, Dr. Sparks and Mr. Soper ask, does a New Zealand variety of parrot called the kea slide down the roofs of alpine huts on its back?

A parrot's foot has opposable toes, two each in front and back, and is the closest birds come to having a hand. With this foot, a parrot can hang onto a branch and eat a seed or nut the way humans eat a sandwich.

And there is, of course, parrot speech.

In captivity, their mimicry flowers naturally. "No formal lessons are required," write Dr. Sparks and Mr. Soper. They recall that Sparkie, a British budgerigar, got himself into the *Guinness Book of World Records* in the 1950's with a virtuoso performance in which he recited eight four-line nursery rhymes without drawing a breath.

The budgerigar, a native of Australia, is the most widespread household parrot in the United States, where it is more familiarly called a parakeet.

Many scientists have shied away from studying parrots. They are hard to study in the wild because tropical foliage screens their activities, and they can be stubborn and cantankerous laboratory subjects. But now, not least because of the endangered status of so many parrots, renewed interest is stimulating research efforts.

In a project that has attracted wide attention, Dr. Irene Pepperberg, an ethologist at the University of Arizona, has been probing the limits of parrot mental ability. "Basically," she said, "we've shown that the parrot is working at the level of the chimpanzee and the dolphin."

Her star subject is a 15-year-old African gray parrot named Alex. In a paper published in 1990 by *The Journal of Comparative Psychology*, she described a study in which Alex was trained to recognize and label objects, colors and shapes, and when questioned, to say their names in English.

Alex was able to identify the shape, color or name of an object correctly about 80 percent of the time. In the pièce de résistance, he was shown a variety of objects—for instance, a purple model truck, a yellow key, a green piece of wood, a blue piece of rawhide, an orange piece of paper, a gray peg and a red box. "What object is green?" he was asked. "Wood," he responded.

In another test, Alex was shown a football-shaped piece of wood, a key with a circular head, a triangular piece of felt, a square piece of rawhide, a five-sided piece of paper, a six-sided piece of modeling compound, and a toy truck.

"What object is five-corner?" the experimenter asked.

"Paper," Alex replied.

Asked 48 such questions, he was right 76 percent of the time—100 percent of the time for questions involving shape. Dr. Pepperberg interprets this as statistically significant evidence that Alex understands the questions as well as the abstract concept of category, and that he thinks about the information to come up with an answer.

In other tests, Dr. Pepperberg said, the parrot has distinguished between the concepts of bigger and smaller, and among the concepts of biggest, smallest and "middlest"—"actually, a difficult concept for children."

"This is not just stimulus-response," she said, pointing out that to answer the questions correctly, Alex must understand them and think about the information. She does not go so far as to say that Alex is using language, but she does describe what is going on as communication between bird and human.

"It's very remarkable stuff, and it is a further indication of how exceptional these birds are in terms of their intellect," said Dr. James Serpell, a zoologist who directs the Companion Animal Research Group at the University of Cambridge department of veterinary medicine in England. "I liken parrots to primates," he said. "They are like monkeys: intelligent, highly manipulative. They use objects almost like tools."

Other scientists urge caution about the Alex experiments, pointing out that they are based on a close, long-term relationship between experimenter and subject and that the experiment is not easily amenable to normal controls and replication. Experts note the difficulty of inferring the thought processes of any animal, whether ape, dolphin or parrot, from its behavior. But, one bird expert who counsels caution, Dr. Fernando Nottebohm of Rockefeller University, said: "That bird is doing some things that look awfully clever and thought-provoking. It does understand questions and gives what seem to be answers."

Many parrots, scientists say, go off the deep end when they are taken from the wild and placed in captivity.

When a parrot is removed from its world and from the tight bond it has formed with its mate, said Dr. Serpell, it is likely to transfer its attachment to a human being. "This is all very well," he said, "but the human being is constantly going away. This is completely unnatural. The parrot would normally never experience that kind of thing."

The resulting frustration, he said, leads to abnormalities like repetitive behavior, in which the bird's head weaves back and forth, or in which it shifts constantly from one foot to the other; abnormal grooming in which the bird plucks out all its feathers, and aggressive behavior. "A lot of them become quite spiteful," said Dr. Serpell. A parrot "might solicit someone to come groom its head and then suddenly reverse itself and bite you," he said.

The wild macaws of Latin America, especially, "make stupid pets," said Dr. Charles A. Munn, a conservation biologist with Wildlife Conservation International, an arm of the New York Zoological Society. "They have tons of neuroses in captivity," said Dr. Munn, who studies wild macaws in the Amazon. "They develop dislikes. They will attach to one owner and hate the other," he said, and become jealous, screeching and yelling, for instance, when a husband and wife sit close together.

Generally, conservationists say, parrots bred in captivity do not display these disturbing quirks. "When they are hand-fed as babies and bond with humans," said Ms. Hemley of Traffic U.S.A., "they do not develop the problems of wild-caught birds." The legislation about to be introduced in Congress would offer incentives to captive breeding programs, which, according to Ms. Hemley, are already expanding.

Overwhelmingly, scientists and conservationists urge people to buy only parrots raised in captive breeding programs and advise them to question pet-store owners carefully. "Ask where the parrot was captive-bred," said Dr. Rosemarie Gnam, a parrot expert at the American Museum of Natural History. "You can check it out by calling up the place." If the store owner hesitates, she said, the buyer should be suspicious. In some states, including New York, it is illegal to sell parrots that come from the wild.

Habitat destruction, trapping and hunting are threatening the existence of 75 parrot species around the world and 40 species in the Western Hemisphere.

Dr. Gnam, who studies the endangered Bahama amazon parrot of the Caribbean, has found that in some years fewer than a third of the eggs hatch successfully. Most are eaten by predators. And Dr. Munn, observing macaws in the Amazon, has found that only 10 to 20 percent of adult macaws attempt to breed in a given year. "These are glacially slow reproductive rates for an animal that's pretty small," he said.

One limitation may be the supply of the holes in trees that parrots require for nests. The macaws seem obsessed with them. "They have a real fixation for cavities," said Dr. Munn. "They can't pass up a good cavity. They're examining things that might be a cavity five years from now." The search for cavities becomes more difficult as deforestation of the tropics proceeds.

Another drag on reproductive success is that the first chick that hatches from the macaw's two-egg clutch gets most of the attention and food, and the second chick frequently dies.

Assiduous efforts are required to gather such data. Dr. Munn, for instance, spends long days in a harness high above the floor of the rain forest, observing macaws through telescopes and binoculars as they gather spectacularly on the bank of a river each day to eat clay. They need it, scientists believe, to neutralize poisons in the seeds they eat.

By watching the daily congregation at the "clay lick," Dr. Munn and colleagues have learned to identify individual birds by their facial markings and observe their relationships and social structure over several seasons. They find that young adults stay with their parents in a tight family unit for several years.

As research goes on, scientists and conservationists are pressing ahead with their plans. Conservationists hope parrots will serve as charismatic "flagship" species whose powerful appeal can be harnessed to the cause of rain forest preservation. In the Bahamas and elsewhere in the Caribbean, there have been education and advertising programs aimed at promoting parrot protection. A broader strategy involves the promotion of parrots as a valuable tourist attraction. One big clay lick in Peru, for example, is near a lodge that has already been established for purposes of what has come to be called ecotourism. So many macaws fly around the lick, and the sight is so brilliantly colorful, spectacular and photogenic, said Dr. Munn, that it is "like Disneyland for parrots."

Native Indians hunt the macaws for food, said Dr. Munn, and he has been trying to convince them that the birds are worth more alive, as a generator of tourist revenue, than dead. One village head man, he said, has expressed support.

"Nothing in the rain forest is worth more than big parrots," Dr. Munn said, "and unfortunately, they are shot and captured in unsustainable ways in most places."

—WILLIAM K. STEVENS, May 1991

Not Just Music, Bird Song Is a Means of Courtship and Defense

Teach me half the gladness

That thy brain must know,

Such harmonious madness,

From my lips would flow,

The world should listen then, as I am listening now.

—*"To a Skylark,"* by Percy Bysshe Shelley

FROM WOODS AND MEADOWS, suburban yards and city parks, a mixed chorus of avian melodies announces the arrival of spring. But many researchers who study bird songs are staying indoors. In laboratories equipped with sound-proof cages, spectrographs, computers and electron microscopes, they are trying to unravel the surprising complexities of these ode-inspiring utterances.

What they are finding suggests that even poets may have underpraised the communicative skills and musicality of songbirds. And students of man may have overrated the uniqueness of human language and the ability to communicate through sound.

"Birds are using the same sensory apparatus we use to process sounds, but they are using it to build a completely different universe than we live in," Dr. Jeffrey Cynx of the Rockefeller University in New York said in an interview. From his studies of bird songs he has come to believe that each animal lives in its own sensory and perceptual world. Thus what we hear in the bird's song may only dimly resemble what the bird hears.

Yet birds' extreme acuity in recognizing the sounds of their own kind helps make them critics, of a kind, of human music. Debra Porter and Dr. Allen Neuringer at Reed College in Portland, Oregon, trained pigeons to differentiate between selections from Bach and Stravinsky. The pigeons were then exposed to music by five other composers, which they had to classify as sounding like Bach or like Stravinsky.

Researchers use a standard technique of behavioral psychology, operant conditioning, to get inside the minds of birds.

First the birds are taught to distinguish between two choices, say, for example, an excerpt from a Bach prelude and a selection from Stravinsky's *Rites of Spring*. Slightly hungry birds are given two disks to peck at. When Bach is played and the bird pecks the left disk, food appears; if the bird pecks the right disk, no food is given and a mild punishment (like the lights going out) may be meted out. The reverse would be used for Stravinsky.

After many trials, the bird "learns" which music is which and makes the correct choice 70 to 100 percent of the time. Then new musical selections can be presented to see how the bird classifies them—as Bach-like or Stravinsky-like—by which disk it pecks at.

The birds correctly classified Buxtehude and Scarlatti as Bach-like and Eliot Carter and Walter Piston as Stravinsky-like. The only "mistake" the birds made, if misjudgment it was, lay in grouping Vivaldi with Stravinsky.

Contrary to popular belief, bird song is not entirely instinctive, although most birds show an innate propensity to learn the song of their species. Songbirds raised in isolation in the laboratory without having heard their species sing develop an incomplete and abnormal version of the song, according to research by Dr. Peter Marler, an animal behaviorist now at the University of California at Davis. But they produce a song that is nearly correct if allowed to hear a tape of it during the so-called sensitive period for song learning, which varies from species to species.

Deaf birds, on the other hand, never come close to singing the right song. Even if they heard it before becoming deaf, they sing with serious distortions, apparently because they cannot hear and correct their own performances.

As for the sequence of song learning, the findings show that birds are very much like human children. As fledglings they babble a so-called subsong of nonsense syllables. As preadolescents they sing often-mispronounced fragments called plastic song. As young adults they are able to articulate properly the song characteristic of their species, crystallized song.

Like children learning to talk, birds learn songs from their elders. But unlike children, who can learn any language they are exposed to, the musical language of most birds is somewhat constrained by their genetic heritage. Given a choice of two songs—their own and that of another, even a closely related species—they will learn their own. But, if exposed only to the song of another species, they will learn a version of it.

Dr. Marler and his colleagues reared male song sparrows and swamp sparrows in isolation. They exposed each to model songs of both species, which have strikingly different songs. But while the song sparrows learned some parts of the swamp sparrow's simple song, the swamp sparrows learned almost none of their close cousin's elaborate melody.

Clearly, Dr. Marler concluded, birds are strongly influenced but not necessarily bound by innate preferences. And, while all birds of a given species may sound alike to a casual birder, researchers have found distinct dialects in different regions. However, songbirds from New York do not seem to have as much difficulty as people do in understanding their confreres from South Carolina, and vice versa.

Birds can even recognize and remember their neighbors' voices. Dr. Renee Godard of the University of North Carolina in Chapel Hill reported recently that male hooded warblers, returning to their breeding grounds after an eight-month absence, recognized the songs of birds that had been their neighbors the previous season. As long as the neighbors were singing from their usual territory, the warblers seemed unconcerned. But if strange neighbors sang, the warblers quickly let them know their proper place. This ability to distinguish friendly neighbors enables the males to spend more time serenading potential mates.

While poets may rhapsodize about the beauty of avian song, the birds themselves seem to have a limited appreciation of their musicality. Dr. Cynx's studies suggest to him that "birds don't really hear melodies—rising and falling pitch—the way we do." If a song the bird knows is transposed up or down an octave, the bird fails to recognize it, his studies showed. In other words, birds respond to absolute pitch (as do only about 5 percent of people) rather than to relative pitch.

On the other hand, he and his colleagues at Rockefeller showed that zebra finches could distinguish subtle differences in timbre or quality of a sound, which is determined by its harmonics. Dr. Cynx told the journal *Bio-Science:* "This is not something you would expect such a small-brained animal to do. It shows an exquisite ability—comparable to that of a skilled musician—to detect, learn, remember and produce the most subtle changes in a complex sound."

When tempos are distorted, Dr. Stewart H. Hulse found, a bird continues to recognize sound patterns even if they are considerably slowed down or speeded up. Dr. Hulse, a psychologist with a musical background at Johns Hopkins University, suggested that "the ability to hear rhythm seems to be present early in animal evolution, but the ability to hear pitch relationships may be unique to humans."

Only male birds sing and only mature birds produce a complete song. Biologists long assumed that the male sex hormone testosterone was responsible for the behaviors both of learning songs and singing. Indeed bird song, Dr. Marler noted, typically "waxes and wanes with the seasonal cycle of testicular growth and regression" and it decreases in castrated birds.

But in studies at Rockefeller University's Field Research Center in Millbrook, New York, Dr. Marler and colleagues showed that castrated birds without testosterone in their blood produced, albeit belatedly, subsong and plastic song. The castrated birds continued to have significant levels of estradiol, a female sex hormone, in their blood during the early stages of song development. The researchers believe it may be this hormone, not testosterone, that "activates the mechanisms of song acquisition during adolescence" and enables birds to memorize songs.

What, then, is the role of testosterone? The brain centers that control the syrinx, the "song box" in the bird's chest, must respond directly to testosterone since the nerve cells have receptors that bind the hormone. Testosterone, then, fuels the motor of song production. It stimulates the growth of cells in these regions of the bird's brain and induces the cells to make more connections.

Dr. Marler, who has overseen decades of studies on how birds learn to communicate, said the research underscores the importance of biology in the development of speech and language. "For too long," he said, "those studying language development in people have overemphasized cultural and social influences at the expense of the biological side."

Dr. Fernando Nottebohm and Dr. Alvarez-Buylla at Rockefeller have studied the brain of a particularly talented songbird, the canary, to demonstrate a phenomenon that may one day help people with brain injuries. Unlike most birds, which cease learning songs after a certain age, canaries can change their song from year to year.

The researchers showed that the acquisition of new songs is accompanied by formation of new cells in a region of the bird's brain called the higher vocal center, which controls song production. Neurons in this center have extensions that contact motor neurons, which in turn activate the muscles of the syrinx where the song is produced.

When female canaries were given testosterone, the Rockefeller scientists found, new connections appeared in the vocal center of their brains. As male canaries dropped old songs and acquired new ones, the birds discarded old brain cells and replaced them with new ones, preventing the adult bird's brain from growing ever bigger.

New cells, the Rockefeller study established, were added to the adult brain in regions other than the song control center as well. While there is still no evidence that new neurons form in the brains of adult primates, Dr. Nottebohm hopes the findings in songbirds may one day help show how to trigger neuron development in people whose brain cells were damaged by disease or injury.

The primary roles of bird song are to define and defend territory and to attract mates. But why, scientists wondered, do some species, like the white-crowned sparrow, have only one song while other species have many? Each song sparrow, for example, sings perhaps 10 songs apiece, marsh wrens and mockingbirds have up to 200 different songs and brown thrashers sing as many as 2,000 songs.

Studies of warblers by Don Kroodsma of the University of Massachusetts in Amherst revealed that each bird had a primary song used to attract a mate, with the remaining secondary songs serving to define and defend territory. Dr. Kroodsma discovered how fast a bird could change its tune when its social situation was altered.

Mimickry and Mozart

I was walking my dog in Prospect Park in Brooklyn one recent morning when I heard a wolf whistle from above. I looked up and saw a starling that had interrupted its apartment house construction to (I presumed) admire my style. I returned the whistled compliment. The starling repeated its call, and I mine. The exchanges might have continued had not my dog objected.

Starlings, though vastly underappreciated by most Americans, were much coveted as pets in Mozart's day. The master himself purchased one after hearing it sing a segment of a concerto he was writing (and had presumedly hummed in the bird's presence).

Birders too have long admired the starling's vocal acrobatics: complex songs and whistles of its own devising and dozens of other utterances acquired from its surroundings, including doorbell chimes and such human utterances as, "This is Mrs. Suthers calling."

One of the most popular mimics in the natural world is the mockingbird, which has fooled many a novice birder with its ability to mimic the songs of other birds. A single male may sing more than 200 different songs, and Merritt and Kim Derrickson of the National Zoo in Washington found that the birds continued to expand their repertoire throughout life.

Dr. Meredith J. West, a psychologist who has studied starling mimickry at Indiana University in Bloomington, believes the behavior functions like "social sonar." The bird does not understand what it is saying, but learns that a particular phrase uttered at a particular time gets a reaction.

The birds rarely reproduce full replicas of human speech or music. Still, Dr. West said, they seem to imitate just the right words and musical fragments to make people like Mozart fall in love with them.

—JANE E. BRODY

Early one morning, he reported in *Natural History* magazine, a recently paired male was busily singing secondary songs until its mate was removed. The next morning the male sang only its primary song—252 times in one hour. A week later he had a new mate and his morning serenade changed to 157 secondary songs and only two primary songs.

Dr. Kroodsma said that many mysteries remain, such as "how the birds know what they are doing and how they develop competence in the use of their vocabularies." He noted that the birds' "use of learned sounds in different situations is intriguingly similar to what we humans do with our language."

Ornithologists at Cornell University, meanwhile, are putting bird sounds to use to protect endangered species. To lure dark-rumped petrels back to their old island haunts in the Galapagos, the scientists planted petrel decoys and played the birds' sounds on solar-powered tape recorders to make the birds think their kind was already there. A similar approach was successfully used by the National Audubon Society to restore puffins and terns to their former homes along the coast of New England.

—JANE E. BRODY, April 1991

Gene May Help Birds Tune Out Sour Notes and Tune In Rivals

MALE SONGBIRDS, those tireless flutists of the forest, must be as adept at appreciating music as they are at making it. A male must sing to seduce a mate and to define the boundaries of his territory; but he must also know the songs of all the birds surrounding him, to distinguish between the harmless tunes of his neighbors and the threatening songs of strangers that may have designs on his little branch of the tree.

Now researchers have discovered the first molecular clue to how male songbirds recognize the melodies of other males. Studying canaries and zebra finches, Dr. Claudio V. Mello and his colleagues at Rockefeller University in New York City have identified a gene that is one of the first to respond in the brains of birds when they hear the songs of other members of their species. The nerve cells that react are in the part of the brain thought to be the avian equivalent of the mammalian auditory cortex, where incoming sound signals are integrated and interpreted.

The scientists do not yet know the purpose of the gene, which goes by the distinctly unmusical acronym of ZENK, but they believe its activation is one of the earliest events in the formation of a permanent memory in the brain. They suspect that once it switches on, the gene sets in motion a molecular program that alters neurons to ensure that the bird learns the tunes of any male it encounters.

The scientists found that the ZENK gene responds most vigorously when the bird hears songs of other males of its species, less robustly when the animal is exposed to the calls of a different songbird species and not at all when a tape of nonsong tones is played. That discrepancy in responses suggests that the gene helps the bird focus on the most important type of sound: the arias of potential competitors.

84

Scientists think they have found a gene that is activated when birds hear songs.

The gene, called ZENK, is most active when a zebra finch hears the song of another finch.

The song of another species sets off a small amount of gene activity in the key brain area.

The gene may start a molecular program that alters neurons to assure that the bird learns the song of any male it encounters.

High ZENK activity.

Patricia J. Wynne

"It is relevant to birds to recognize other birds' songs," said Dr. Mello. "These are territorial males that will respond with aggressive, attacking behavior if a new bird intrudes, and they localize that intruder by hearing its song."

The new experiment is part of a growing effort among biologists to bridge the canyon between microscope and macroscope. Scientists generally study either the behavior of animals in the wild or the details of molecules in the test tube. Dr. Mello's research is an attempt to apply the precision of molecular biology to the somewhat scruffier world of bird behavior. And while other researchers have studied the neurobiology of bird song and have mapped centers of the brain that are critical to singing, the latest report goes further, describing the individual genes at work within those neural domains.

The report appears in *The Proceedings of the National Academy of Sciences.*

"This is one of the first links I'm aware of between research on a natural learning process and the switching on" of a gene, said Dr. Peter Marler of the University of California at Davis, an authority on the behavior of bird song. "It's probably not an exaggeration to call this the dawning of a new era in research on the development of behavior."

Dr. Marler suggests that the ZENK gene may have its human equivalent, and that it may play a role in the ability of a baby to begin mastering language, but he admits that this idea remains highly speculative.

In the new work, the Rockefeller researchers chose to examine the ZENK gene in bird brains because it had been shown through other experiments to participate in brain activity and to respond to changes in stimulation. For example, when researchers disturbed the circadian rhythms of rodents by altering their exposure to light and darkness, the ZENK gene blazed to life in the brain region in charge of the biological clock.

To look at the pattern of gene activity in birds, the scientists took a total of 24 adult male canaries and zebra finches, and kept the birds in isolation for 24 hours. Each male was then put in a box and exposed for 45 minutes to a tape recording of a same-species song, another species' song or simple tones. During each session, the bird would sit quietly and in apparent concentration, as birds normally do upon hearing the chirrups of other males. Afterward, the bird was sacrificed, and its brain was cut into sections about one cell thick.

Using a radioactive probe corresponding to the ZENK gene, the researchers looked for evidence that the gene was active, or expressed, in different parts of the brain. They found that gene expression was most pronounced in those birds that had heard their compatriots' songs, and that the region involved was the caudal neostriatum, a part of the bird's forebrain where signals from the ears are interpreted. Dr. Mello said he believed that the arousal of the gene probably began quite early in the listening exercise, perhaps in fewer than 10 minutes.

Researchers also looked at the brains of female songbirds, which do not sing. They nevertheless found gene activity in the forebrain, indicating that the females needed the gene to recognize the songs of their own species.

What happens after the initial gene burst remains to be determined. The ZENK gene is thought to control the activity of other genes, to mobilize them as part of a significant alteration of the neurons and the creation of long-term memory. Auditory information from the forebrain may also travel to the bird's higher vocal center, the region of the brain in charge of the animal's own song production—in essence, allowing the bird to plagiarize the riffs of others to compose its own mating melody.

—NATALIE ANGIER, August 1992

4

BIRD BIOLOGY AND BEHAVIOR

New studies of bird behavior are bringing to light some unexpected complexities. Ever wonder why magpies allow cuckoos to stick them with the cost of raising cuckoo offspring in their nests? The cuckoos hang around to enforce the adoptions, killing the magpie chicks in nests where the parents disposed of their cuckoo egg gift.

A study of how female barn swallows choose their mates has shown that they prize two qualities in a male's tail: length and symmetry. Why symmetry? It turns out that symmetry is a fairly accurate measure of how heavily infected the male is with parasites; this in turn is a measure of his general state of health.

The Seychelles warbler can apparently determine the sex of its chicks, producing more males or females depending on the sex ratio required by environmental conditions. It is not yet understood how the trick is accomplished.

Birds in the course of their long evolution have accumulated a rich repertoire of behaviors. Biologists exploring this treasure chest are probably still nearer to the top than the bottom.

Thuggish Cuckoos Use Muscle to Run Egg Protection Racket

BIOLOGISTS had ranked them among nature's most laughable dupes, inexplicably gullible "birdbrains" that dutifully tended eggs dumped into their nests by other bird species. For evolutionary biologists, the many species of birds that so devote themselves to a stranger's young have been something of a mystery, for even when the dumped eggs and young look nothing like their own, the birds often favor the parasites' offspring at the expense of their own.

Now a study in the journal *Evolution* offers the first evidence to support what had been considered an unlikely explanation for this behavior. Biologists studying magpies and the great spotted cuckoos that dump eggs into their nests say that the magpie hosts are not dupes at all, but have been forced into cooperation by an avian extortion scheme.

The researchers say the cuckoos return periodically to check on the nests in which they have left their eggs. If they find their young safely there, all is well. If their eggs are missing, tossed out by uncooperative magpie hosts, the cuckoos destroy the nest, killing the remaining egg or chick inhabitants wholesale. In other words, the cuckoos are members of an avian mafia.

"It's an offer that the birds cannot refuse," said Dr. Anders Moller, an evolutionary biologist at Copenhagen University in Denmark and an author of the study. "It's just the same as in the human mafia. If you resist, it turns out very badly."

Dr. Timothy Clutton-Brock, an evolutionary biologist at Cambridge University in England, called the paper "extremely interesting," saying that such punishment behaviors were probably widespread among animals for keeping others in line. He describes this apparently reliable and adaptive

Michael Rothman

One Pushed Out of the Magpie's Nest

When the great spotted cuckoo ejects an egg in the process of destroying a host's brood, it may be exacting punishment for the bird's ejection of a cuckoo egg. In an effort to keep its own egg safe until it is hatched by the unwilling host, the cuckoo returns periodically to check on the nest.

strategy for living as: "You do something nasty to me, I do something even nastier to you."

Raising a nest full of eggs and chicks is difficult, time-consuming work. There is the incubating of eggs, the chasing off of predators, the finding of food for so many peeping, gaping mouths, not to mention feeding oneself to maintain the energy to do all this intensive baby-rearing. So cuckoos might well be expected to have evolved all manner of tricks to get other birds to do such work for them.

But Dr. Manuel Soler of the University of Granada in Spain said that he and his colleagues did not believe that birds engaged in such coercive behavior and had set out to disprove the theory known as the mafia hypothesis. Dr. Soler studied the great spotted cuckoos and the magpies they parasitize in high altitude plateaus in southern Spain. He worked with his brother, Dr. Juan Soler, and Dr. Juan Martínez, behavioral ecologists at the university, and Dr. Moller.

To test the hypothesis, Dr. Soler and his colleagues removed cuckoo eggs from 29 nests while leaving them in 28 nests. What they found was that in most of the nests that had had their cuckoo eggs removed either the magpie eggs or chicks that remained were later killed. In contrast, nearly all the nests in which scientists allowed the cuckoo eggs to remain were left intact.

At the same time, scientists monitored nature. The great majority of nests from which magpies had ejected cuckoo eggs on their own, without the help of scientists, were also attacked and their young inhabitants killed. Very few of those magpie nests that accepted the cuckoo eggs suffered such attacks.

Such killings, like most rare and rapid events in nature, are hard to witness. But the biologists say they are confident that the attackers were indeed the cuckoos whose eggs had been ejected. When removing eggs from nests to set up their experiment, the researchers were often scolded by cuckoos, which quickly checked the nests after researchers were done. They also followed one female cuckoo outfitted with a radio transmitter who returned to a nest from which her egg had been removed and destroyed the contents.

But most convincing was the evidence in the nests themselves. For what the biologists found were pecked eggs and wounded nestlings, all left behind by their killers. While other birds and animals attack magpie nests, such hungry predators do not leave their victims behind.

By the breeding season's end, the magpies that accepted cuckoos in their nests tended to produce more magpie young than those that ejected them, suggesting that the cost of noncompliance is high.

"The experiment they did is very convincing," said Dr. Peter Arcese, an ecologist at the University of Wisconsin in Madison. "People are going to have to take seriously the idea that these nest parasites are more sophisticated than we think."

Researchers say the data are the first to support the so-called mafia hypothesis proposed in 1979 by Dr. Amotz Zahavi, a behavioral ecologist at Tel Aviv University in Israel. Dr. Zahavi proposed that nest parasites, like the cuckoo, might be bullying their hosts into accepting eggs under threat of violence if they did not. But in the 16 years since Dr. Zahavi's hypothesis was published, no evidence had turned up in support of it.

"He's put out a number of ideas that people have initially pooh-poohed," said Dr. Arcese, "and later people have shown that, in fact, they may operate."

Dr. Zahavi said, "Obviously it is satisfying that a model you created is found to be true at least for one cuckoo in one place."

But at the same time, researchers note that enforcement may not be the only reason that parasites like the cuckoos are destroying nests.

Dr. Arcese said that based on studies of cowbirds that parasitize song sparrows on Mandarte Island near Victoria, British Columbia, he and his colleagues had evidence that cowbirds could also cause their hosts' nests to fail. But Dr. Arcese says their studies indicate that the cowbirds may be destroying nests, not to teach the song sparrows a lesson, but for their own convenience.

Cowbirds, like other nest parasites, must find nests into which eggs are being freshly laid. In nests with older eggs or eggs of unknown age, the host's young may hatch first, ending incubation and leading to the death of the parasite's egg.

To avoid such problems, Dr. Arcese suggests that parasites, including the cuckoo, may kill young as a way of getting hosts to start another nest, where the parasites can leave their eggs at the perfect time.

Dr. Stephen Rothstein, an evolutionary biologist at the University of California at Santa Barbara, while praising the team's work as "superb," suggested a simpler explanation for the fact that many magpies keep the cuckoo eggs.

While the eggs and young of many parasites look strikingly different from that of their hosts, those of the great spotted cuckoo are good mimics of the magpie's.

"It could just be evolutionary lag," said Dr. Rothstein, describing an idea that has come out of his work with cowbirds. That is, magpies may keep cuckoo eggs simply because they have not yet evolved the ability to make the sometimes difficult distinction between the cuckoo's and their own. It is a lag that leaves the cuckoos winning the evolutionary war, at least for now.

Dr. Rothstein added that he also had evidence that parents of nests from which any eggs had been removed, whether the bird's own or a parasite's, would often desert the nest. He said this could explain the greater rate of attacks on nests from which eggs had been experimentally ejected as seen in the new study. With eggs missing, the magpie parents might be considerably less interested in tending and protecting the nests, leaving them open to attack by cuckoos or other birds.

To complicate matters even further, Dr. Rothstein said he and his colleagues have studied the same parasite, the great spotted cuckoo, in Israel where it leaves its eggs in crows' nests. Doing similar experiments, they found no evidence of mafia behavior.

But Dr. Arcese said that more and more researchers seemed to be finding such geographical differences in the behavior of these birds. One explanation is that since both the parasites and their hosts are long-lived and can learn, these complex behaviors may actually differ from place to place, depending on what they have experienced.

At the same time, researchers say that both the great spotted cuckoo and the cowbird are extending their ranges, moving into new territory and encountering new birds. Biologists say that with such changes going on, rather than some studies being wrong, all may be right, with researchers witnessing different stages in the ongoing skirmishes of the evolutionary war between these parasites and their hosts.

—CAROL KAESUK YOON, November 1995

For Remote Bird Species, the Sex of Hatchlings Is No Surprise

AMONG PARENTS, one question never fails to arouse a lively debate: Which is the preferable sex to raise, boys or girls? While nearly everyone has an opinion on this impossible-to-resolve question, it remains the uncontrollable biological reality of sperm meeting egg rather than parental preference that casts the die for a little Jack or tiny Jill.

One thousand miles off the east coast of Africa, however, on a remote island in the Seychelles, researchers have discovered a bird that is able to produce a male or a female offspring according to its changing needs with uncanny precision.

"I was surprised," said Dr. Jan Komdeur, a lecturer in animal behavior at the University of Melbourne in Australia, who discovered the bird's behavior. "But I thought it was a very clever thing." Dr. Komdeur and his colleagues published their study in the journal *Nature*.

In fact, evolutionary theorists have been arguing since early in the century that ideally, parents of any species should adjust the ratio of male to female offspring that they produce, depending upon the evolutionary payoff, or how much it costs to produce a male or female relative to how many young that offspring is likely to produce. But for many years, only wasps and bees showed any indication that they could alter the ratio of male to female young.

Now new studies, including the latest and most striking one with Seychelles warblers, are beginning to show that whether they be rats, birds or bees, when it comes to producing young, parents can sexually discriminate and do.

"This is very big news," said Dr. Patricia Adair Gowaty, a biologist at the University of Georgia in Athens, calling the study an "instant classic."

Dr. Stephen Emlen, the Jacob Gould Schurman Professor of Behavioral Ecology at Cornell University in Ithaca, New York, called the study "awesome," saying: "It's going to raise eyebrows. There are going to be people so astonished by this that they are going to challenge it."

In the Seychelles, the only place in the world where these warblers are found, they live in pairs, each on their own territory. But on the 70 acres of tiny Cousin Island, there are not always enough territories for every bird to get a mate and make a home. So many young birds, usually females, stay at home with their parents and learn the tricks of the trade by helping out. They help build nests, defend the territory, incubate eggs and feed newborn chicks.

Helpers are only beneficial when there is enough food on the territory and no more than two such assistants. With more helpers or too little food, they become a liability as they compete with parents and chicks for scarce insect meals. So, the theory goes, parents should produce females when helpers would be useful and males when they would not.

Using DNA tests to determine the sex of newly hatched chicks (both sexes look the same), researchers found that the birds produce young in their natural habitats exactly as theory would predict.

Researchers were also able to carry out an unusually convincing test. They moved pairs of warblers from territories without much food on Cousin to territories with lots of food on the neighboring islands of Aride and Cousine (not to be confused with Cousin, the island of origin) where no Seychelles warblers live. Pairs that had produced mostly or all males on Cousin when they had little food to share suddenly began producing mostly or all females to help out in the more bountiful new territories.

A huge obstacle in studying birds had been the difficulty of telling the sex of chicks. But even in those few species where researchers could tell male and female chicks apart, the deviations from a 1-to-1 sex ratio were so subtle that many doubted if parents were up to anything at all. In fact, the conventional wisdom became that birds were physiologically incapable of any precise control over what sex chick they produced.

Bees and wasps, on the other hand, have long been thought to manipulate the sex of their young.

For example, Dr. John Werren, an evolutionary geneticist at the University of Rochester, studies parasitic wasps that lay their eggs on caterpillars and other insects. Large female wasps that have grown up on a big juicy caterpillar do much better than small females, as they are able to produce many more eggs. Males, on the other hand, do fine whether they are large or small. What researchers have found is that quite logically, when a mother wasp finds a nice big caterpillar, she lays a female. When she finds a puny one, she leaves behind a male.

But in wasps and bees, the method for manipulating the sex of offspring is clear. When a mother wasp wants to produce a male, she lays an unfertilized egg. To produce a female, she fertilizes the egg with her cache of sperm before laying it.

But as for the birds, "how in heaven's name are they doing it?" Dr. Emlen asked. "No one has any idea."

In birds, sex is determined, as in humans, by sex chromosomes. In humans, men carry two different sex chromosomes, an X and a Y, and women carry two X chromosomes. That means every woman's egg has an X chromosome, and the child's sex is determined by the father's sperm, which provides either a Y, for a boy, or an X, for a girl.

In contrast, in birds it is the females that carry the two different sex chromosomes, a Z and a W, while male birds carry two W chromosomes. So it is the female bird and her egg, carrying a Z or a W chromosome, that determines a chick's sex.

While noting that it remains to be seen just how these birds do it, in principle, Dr. Werren said, a bird that can watch her Z's and W's can exercise a lot of potential control over the sex of her young.

The evolution of sex ratios may appear to be an esoteric topic, but Dr. Gowaty said it might be closer to home than it seemed. "One exciting aspect is if we could select for lines of all-daughter or all-son producers, for example, in chicken or in cattle," Dr. Gowaty said. "You know it's been tried for years and years. But you haven't heard about it, right? That's because no one's been able to do it. But wouldn't that be economically wonderful?"

While the Seychelles warblers, also known as *Acrocephalus sechellensis*, are a far cry from an all-hen-producing chicken, they are an indication that at least some birds have the capacity to produce all or mostly one sex of young under the right conditions.

Dr. Martha K. McClintock, a biopsychologist at the University of Chicago, said the new sex ratio studies might hit close to home in quite another way.

While there is nothing quite so dramatic as the switches seen in Seychelles warblers, evidence is growing in mammals that mothers tend to produce more male offspring when males may be most successful in mating and more females when they are the better bet for producing new generations. That work, she says, speaks to humans as well.

In mammals, in general, the main method for controlling the sex of the offspring appears to be loss: loss of sperm not used in fertilization and loss of fetuses before they are born. What determines the sex ratio, for example in a litter of pups, is the pattern of loss along the way.

Referring to Dr. Leslie Hornig, also at the University of Chicago, Dr. McClintock points out that there are similar patterns of loss in humans. "We all know it's the sperm that determines the sex, but as Leslie points out, there are millions that don't make it. What determines the ones that do?" Likewise, Dr. McClintock said, the natural abortion of fetuses in women happens more often with male than female fetuses, suggesting that women's bodies may be selecting for or against fetuses of different sexes as well.

So while men's sperm have long borne the sole responsibility, women's bodies may have much more control than previously imagined over which sex is ultimately born.

Dr. McClintock summed it up, saying, "Henry the Eighth might have been more right than we give him credit for."

—CAROL KAESUK YOON, February 1997

Seeking Food, Some Sea Birds Rely on the Sense of Smell

THE IMPORTANCE of the sense of smell for most birds is a huge question mark, but experiments in the South Atlantic suggest that some sea birds, especially storm petrels, use smell to find food and perhaps navigate in the vast and seemingly featureless ocean.

The sea birds' apparent cue is the compound dimethyl sulfide (DMS), produced when drifting plant plankton, or phytoplankton, are being grazed upon by shrimp-like crustaceans called krill and by other minute animals called zooplankton. Krill are the main food of many sea birds, including penguins.

Scientists aboard a British Antarctic Survey ship, the *James Clark Ross,* found that Wilson's and black-bellied storm petrels, white-chinned petrels and prions were consistently attracted to oil slicks laced with dimethyl sulfide. Dr. Gabrielle Nevitt, a neurobiologist at the University of California at Davis, who led the research team, said these birds often forage at night when olfactory cues would be particularly valuable. "This constitutes the first evidence that dimethyl sulfide is part of the natural olfactory landscape overlying the southern oceans," she said.

Dr. Nevitt described the chemical compound as "smelling like seaweed rotting on a beach," adding, "In high concentrations it's fairly repulsive."

Large albatrosses and cape petrels, on the other hand, did not respond to dimethyl sulfide slicks any more than to control slicks of vegetable oil. These sea birds forage by spotting and exploiting feeding aggregations of seals, whales and other conspicuous sea birds, Dr. Nevitt wrote in describing the findings in the journal *Nature*. Her colleagues in the study were Dr. Richard Veit and Dr. Peter Kareiva, both of the University of Washington in Seattle.

Although birds have well-developed olfactory organs, relatively few studies have been done to determine how they use the sense of smell, if at all. "Smell has been shown to be important in the lives of only a few bird species," said Dr. Christopher Duncan, a professor of zoology at the University of Liverpool in England, an expert on the topic. "This is an extremely interesting experiment."

Soaring turkey vultures are known to locate dead animals by smell, although other New World vultures apparently depend on eyesight to find carcasses. The flightless kiwi, a nocturnal and nearly blind inhabitant of New Zealand forests, detects worms and insect larvae in the leaf litter by smell. The kiwi's nostrils are near the tip of the bill rather than at the base, as in most other birds.

Petrels, prions, storm petrels, shearwaters and albatrosses comprise a large order of oceanic birds called tubenoses because their nostrils are located in conspicuous, horny sheaths at the base of the bill. Also, their brains have large olfactory lobes and scientists have long suspected that these species hunt by smell. Earlier trials showed that they would follow odor trails from sponges soaked in cod liver oil.

"Wilson's storm petrels are like bloodhounds," Dr. Nevitt said. "They flew into DMS slicks more than twice as often as into control slicks." Smaller than a robin, Wilson's storm petrel is one of the world's most abundant birds and while it breeds on the Antarctic shore, it is commonly seen off the east coast of North America. Sooty brown with a white rump patch, Wilson's storm petrel patters along the wave tops, plucking plankton off the ocean's surface.

Prions are somewhat larger petrels and 22 million pairs nest on South Georgia Island in the sub-Antarctic waters where the experiments were done. The white-chinned petrel is a heavy bird with a four-foot wingspan that nests alongside albatrosses on the tussocky slopes of the southernmost islands.

"Our aim was to produce a downwind DMS concentration similar to natural levels that a petrel would encounter," Dr. Nevitt said. The compound is a byproduct of metabolic decomposition in marine phytoplankton, and the process is sharply accelerated during grazing by zooplankton. "The emissions can persist for hours or even several days," she said, "and DMS is a major source of atmospheric sulfur."

Plankton and dimethyl sulfide concentrations may be signposts for tubenose sea birds that spend most of the year roaming the oceans on transequatorial migrations.

"DMS concentrations tend to be highest in surface sea water over shelves and in upwelling zones where nutrients rise to the surface," Dr. Nevitt said. "These oceanic features, reflected in the atmosphere as elevated DMS profiles, could form an olfactory map for migrating sea birds."

—LES LINE, August 1995

Raptors Found to Track Prey's Ultraviolet Trail

BECAUSE KESTRELS, as they hover above open land, can see in the ultraviolet light range, they are apparently able to spot the highways of voles and other small mammals and swoop on their prey, according to a Finnish study.

The birds can see the trails from far overhead because the voles, like other mammals such as mice, dogs and wolves, mark their trails and territory with urine or feces. The waste material is not only tagged with the odor of their species but also marks their highways because it absorbs ultraviolet light.

Many raptors, such as falcons and sparrow hawks, the American counterpart of the kestrel, are believed to have ultraviolet vision and, in treeless regions, this could be an aid to hunting.

In Scandinavia and other northern regions the populations of small mammals often oscillate in four-year cycles between overabundance and crashes. In a crash it becomes hard for raptors, such as kestrels and snowy owls, to find their prey. If, however, they can see a highway map of their prey's movements, they can rapidly scout large areas.

The kestrel is a small falcon noted for hovering upwind while turning its head, hunting for prey. As the Finnish researchers pointed out in the journal *Nature,* crashes in vole populations can affect very large areas, sending kestrels on searching expeditions of many hundreds of miles. With ultraviolet vision, however, they can spot surviving colonies of their prey.

This is particularly true in the spring, before grass has begun to hide the runways. The authors wrote, "We have provided the first experimental evidence, to our knowledge, of a wild raptor using vole trail marks to select hunting patches and potential nest sites." Other small mammals, such as mice, are known to produce urine that fluoresces in blue light.

The Finns tested captured wild kestrels at the Konnevesi Research Station 150 miles north of Helsinki and in level farmland at the nearby Alajoki study area. The land was plowed each autumn, leaving the voles no place to live except in ditches. With most of the landscape devoid of voles it was suitable for the creation of artificial vole trails, and it was found that the kestrels favored the trails that were illuminated by ultraviolet light.

—WALTER SULLIVAN, April 1995

Eggs on Feet and Far from Shelter, Male Penguins Do a Shuffle

Michael Rothman

EVER SINCE EXPLORERS at the turn of the century surmised that the emperor penguin must breed during the coldest part of Antarctica's deadly winter, this charming but peculiar species has inspired wonder and curiosity. But after 83 years of penguin watching, scientists are still encountering surprises, including a recent discovery made with a night-vision device that revealed emperor penguins to be even hardier than had been supposed.

Knowledge of the winter behavior of emperor penguins has come slowly because it is hard for human beings to survive under such conditions, much less conduct scientific research. High winds, temperatures around minus 60 degrees Fahrenheit, round-the-clock darkness and frequent sudden blizzards make the Antarctic habitat of the emperors, the Ross Sea and McMurdo Sound, dangerous places in winter.

In 1911 a trio of intrepid English explorers confirmed the theory that emperor penguins, unlike other birds, do indeed lay and incubate their eggs during the dead of the polar winter. Two of these brave investigators, Dr. Edward A. Wilson and Henry Bowers, were doomed to die of starvation and exposure with Capt. Robert F. Scott a year later while returning from their ill-fated expedition to the South Pole.

Risking their lives in the darkness by climbing down a 200-foot cliff in a bitterly cold wind, Dr. Wilson, Mr. Bowers and Apsley Cherry-Garrard succeeded in reaching a mid-winter emperor penguin rookery at Cape Crozier on Ross Island. Collecting five penguin eggs with very young embryos for later studies of biological development, the explorers noticed that the big birds had chosen a spot offering substantial shelter from the wind.

Ever since then, scientists had assumed that all emperor penguin rookeries must be similarly placed, so a recent discovery by Dr. Gerald L. Kooyman of the Scripps Institution of Oceanography in San Diego came as a surprise. During a nonstop jet flight last June from New Zealand to the South Pole and back, Dr. Kooyman scanned the darkened expanse of sea ice near Ross Island using an image intensifier of the kind supplied to American forces in the Vietnam and Persian Gulf wars.

For human beings, Antarctica is closed to the outer world during the winter months, from March through August, and only in dire emergencies do airplanes from New Zealand or Chile attempt to land on the continent's treacherous ice runways. But each winter the National Science Foundation arranges one flight from New Zealand to airdrop fresh vegetables and mail to the inhabitants of McMurdo and South Pole Stations. Dr. Kooyman took advantage of this winter's 13-hour flight to spy on the emperors.

From his perch aboard a KC-10 tanker aircraft 10,000 feet above the frozen sea, Dr. Kooyman spotted a cluster of about 7,000 emperor penguins huddled together on sea ice west of Ross Island, a considerable distance from the shelter of cliffs. It was instantly clear that emperor penguins can breed even in the absence of any kind of shelter.

"They were all males, of course," he said, "and each one was presumably incubating an egg." Male emperor penguins are known to take over the incubation of eggs for nine weeks, going without food the whole time while their mates return to the open sea to fatten up on deep-dwelling fish.

The emperor penguins' strategy for keeping warm in winds of minus 60 degrees Fahrenheit, Dr. Kooyman said, is to keep shuffling around and jostling their neighbors, taking care to keep their big eggs wedged between the tops of their feet and their warm, overhanging bellies. The penguins at the exposed periphery of a cluster generally manage to work their way inside after a fairly brief period acting as community windshields.

Scientists do not know how long an emperor usually remains at the exposed edge of a community "huddle," or how rapidly the entire huddle moves. Weather conditions and the stability and thickness of the sea ice on which they stand are believed to be factors.

"So few of these birds have ever been tagged that we have little knowledge of their individual movements," Dr. Kooyman said. "We suspect that movement is weather-dependent; in mild weather with low winds, the huddle probably loosens up, while high winds and low temperatures draw the birds together and increase the speed of their movements. There may also be a learning curve. It may take inexperienced birds a while to learn how to jockey for position."

Both male and female emperor penguins protect their eggs and newly hatched chicks by enveloping them under a fold of body skin. During the first part of the reproductive cycle the mothers must fast, but after eggs are laid, they leave to fatten up. The males then take over, incubating the eggs and the newly hatched chicks for the nine mid-winter weeks. The part of the bird's belly in contact with the egg is bare of feathers to facilitate the flow of heat from the father to his offspring. At the end of their baby-sitting stint, the fathers turn the chicks over to their returning mothers, and go back to sea.

If weather conditions become so severe that a parent's resources can no longer cope with the cold, it abandons the egg to save itself, as it may do under certain other circumstances. All penguin rookeries are littered with abandoned eggs and dead chicks.

In many other ways as well the emperor penguin differs from about 16 other penguin species, whose ranges extend from Antarctica to the equatorial Galapagos Islands. Standing sometimes nearly 4 feet high and weighing some 90 pounds, the emperors are the largest of all penguins. Emperors and Adélie penguins are the only two penguin species inhabiting the Antarctic continent proper, although other species live among the northern islands of Antarctica.

Adélies breed in the Antarctic spring on densely populated rocky rookeries, and before the summer sun sets in March, at the beginning of the long winter night, the chicks are mature enough to look after themselves. But as early as 1911, investigators realized that emperor penguins had to contend with an unusual reproductive problem: their chicks mature too slowly to permit a normal avian breeding cycle.

In his 1923 memoir, *The Worst Journey in the World,* Mr. Cherry-Garrard recounted the discovery that if emperor penguins had to begin breeding in springtime like the Adélies, their chicks would still be immature and lacking sufficient feathers and body fat by the time winter set in.

Emperors have some other characteristics that intrigue scientists, including their ability to hunt for prey at great depths. "We've measured emperor dives to depths as great as 630 meters," or 2,070 feet, Dr. Kooyman said. "They are surely the deepest diving birds in the world."

Dr. Kooyman plans to return to Antarctica next month to study the birds at close hand during the summer. The penguins themselves seem to welcome human visitors. Although they are frightened by helicopters flying overhead, a helicopter that lands on the sea ice quickly draws crowds of the birds, all honking enthusiastically and waddling up as fast as their feet can carry them.

Scientists regard Antarctica as a kind of gauge for measuring the health of the global environment, and one of the indicators is the waxing or waning of animal populations, including those of penguins. Among the potentially adverse effects on the Antarctic environment is the impact of increasing tourism.

But Dr. William Fraser of Montana State University recently reported an anomaly in penguin colonies that remains to be explained. Two islands near Palmer Station on the Antarctic Peninsula, Torgerson and Litchfield, both have large populations of Adélie penguins. Tourists are allowed to visit

Torgersen, and even nesting penguins have become accustomed to having people around. But Litchfield is strictly off-limits. Although Litchfield and Torgersen are right next to each other, the Adélie population of protected Litchfield has declined 43 percent since 1992, while the population of Torgersen fell only 19 percent.

The reason, Dr. Fraser said, apparently has nothing to do with tourists, but is the result of weather. For the last 20 years, he said, the region near Palmer Station has been steadily warming, and this has resulted in increased snowfall every winter. But the snow does not accumulate evenly; the prevailing storm winds sweeping the Antarctic Peninsula come from the northeast, and snow tends to pile up on the lee side southwest of hills and rocks.

Torgerson Island is fairly flat and snow is quickly swept away, but Litchfield has many rocky pinnacles and obstructions, behind which snow accumulates to depths of three feet or so. Adélie penguins are programmed by nature to follow a tight reproductive schedule, and the peak of their egg-laying period is 10 days, from November 18 to 28. If snow deeply covers their rookery during that period, high infant mortality results.

A continuous warming trend on the Antarctic Peninsula has been measured since 1974, but Dr. Fraser believes that the Adélie population has been declining for as long as 50 years because of warming and consequent changes in weather patterns. Few scientists believe there are enough data yet to prove that global warming is responsible, but specialists are keeping a close watch on Antarctica.

"The answer requires the long-term data we're now collecting," Dr. Fraser said.

—MALCOLM W. BROWNE, September 1994

Juvenile Penguins Go Beyond Safe Area

USING SATELLITE TELEMETRY, scientists have tracked juvenile emperor penguins from their natal colony at the Ross Sea in Antarctica to distant ice-free waters in the Southern Ocean, outside the area where they are protected under international agreements.

"It is disturbing to learn that emperor penguins leave the relative safety of the Ross Sea during at least one critical stage in their life cycle and range in areas that are, and will become, more heavily exploited by commercial fisheries," said Dr. Gerald L. Kooyman, a researcher at the Scripps Institution of Oceanography in San Diego.

Dr. Kooyman expressed concern that large numbers of the penguins could be caught in the nets or long lines of fishing fleets, and that stocks of the marine animals on which they feed could be overfished.

The provisions of the Antarctic Treaty and the Convention on the Conservation of Antarctic Marine Living Resources extend only to the 60th parallel in the Ross Sea region, Dr. Kooyman pointed out in a report in the journal *Nature*. But four juvenile birds from an emperor penguin colony near Cape Washington were tracked well beyond that boundary, and one penguin was 1,775 miles from its birthplace when signals from its transmitter stopped at 56.9 degrees south latitude.

Dr. Kooyman said these findings suggested that the 60th parallel was "too limiting to adequately protect even the most familiar symbol of Antarctic wildlife, the emperor penguin." He urged that scientists investigate the dispersal patterns of Adélie and chinstrap penguins, which also breed on the coast of Antarctica.

The satellite tracking of young emperor penguins, using small transmitters glued to the feathers on their backs, is part of a continuing study of the species by Scripps scientists. The Ross Sea birds, Dr. Kooyman said, "are of special interest because they occur in the last marine frontier that has not been exploited by humans."

The geography of the Ross Sea, he added, provides the penguins with a safe haven compared with birds from breeding sites on the Antarctic continent's outer perimeter. About two dozen emperor penguin colonies are known and the species' breeding population is estimated at 135,000 to 175,000 pairs. The bird's numbers have declined at some colonies in recent years and disturbance from human activity, particularly helicopter flights, at nearby scientific bases, is cited as one likely cause.

The entire breeding cycle consumes nine months. The parent birds depart to put on weight and molt shortly before their chicks fledge, leaving the 25-pound juveniles to fend for themselves. Dr. Kooyman said the adult emperor penguins travel in ice-congested polar waters during the non-breeding months, feeding on fish and squid at depths up to 1,500 feet. But the whereabouts of the juveniles and their food habits from the time they leave the colony until their return several years later are still a mystery.

Because transmitters and antennas were attached near the penguins' tails to reduce drag, satellite reception was possible only when the birds were out of the water, resting on pack ice or a drifting iceberg. The juvenile emperors left the Ross Sea colony in late December, and the researchers expected they would remain in the pack ice like the adults. The transmitter batteries would have lasted at least through June, but the last signal was received on March 6. By then, Dr. Kooyman said, the birds were in ice-free waters, where they were immersed at all times, traveling at an average rate of one mile an hour on a course that would have them circumnavigating Antarctica for the next few years.

Juvenile emperors, he added, are not as physiologically capable as the adults, and the availability of prey that can be caught in shallower dives "may be the key as to why they go so far north."

—LES LINE, October 1996

Why Birds and Bees, Too, Like Good Looks

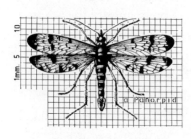

Barn swallow tail
Female barn swallows prefer a long-tailed male with a symmetrical wishbone pattern of feathers the same size and color on both sides of the tail.

Michael Rothman

BEAUTY IS ONLY SKIN DEEP. How sweet that old chestnut is, equally comforting to the unbeautiful, who know they have so much beyond physical appearance to offer the world, and the beautiful, who, after years of being pursued for their prettiness, really do want to be loved for their inner selves.

The only problem with the cliché, say evolutionary biologists, is that it may not be true. In the view of a growing number of researchers who study why animals are attracted to each other, a beautiful face and figure may be alluring not for whimsical esthetic reasons, but because outward beauty is a reasonably reliable indicator of underlying quality. These biologists have gathered evidence from studies of species as diverse as zebra finches, scorpion flies, elk and human beings that creatures appraise the overall worthiness of a potential mate by looking for at least one classic benchmark of beauty: symmetry.

By this theory, the choosier partner in a pair—usually, though not always, the female—seeks in a suitor the maximum possible balance between the left and right halves of the body. She looks for signs of exquisite harmony, checking that the left wing is the same length and shape as the right, for example, or that the lips extend out in mirror-image curves from the center of the face. In searching for symmetry, she gains essential clues to the state of the male's health, the vigor of his immune system, the ability of his genes to have withstood the tribulations of the environment as he was growing up.

The new emphasis on the importance of symmetry to mate choice is one of those annoying developments in evolutionary research that lends oblique validity to ingrained prejudices—in this case, to a fairy-tale view of the world, in which princes and princesses are righteous, strong and lovely, while the bad folk are misshapen and ugly. Biologists emphasize that symmetry is just part of the story of how animals make their choices, and that much remains to be learned about what, in any given species, the possession of a perfectly proportioned body announces to one's peers.

Nevertheless, symmetry does seem to play a role in desirability. Reporting in the journal *Nature,* Dr. John P. Swaddle and Dr. Innes C. Cuthill of the University of Bristol in England found that when they put a variety of colored bands on the legs of male zebra finches, the females vastly preferred males with symmetrically banded legs over those given bands of different colors on each leg, a manipulation that apparently made the males look as goofy to potential mates as somebody wearing mismatched socks.

Writing in the journal *Trends in Ecology and Evolution,* Dr. Paul J. Watson and Dr. Randy Thornhill of the University of New Mexico in Albuquerque sum up the data gathered thus far on the role of symmetry in mate selection. In their own work, they have shown that female scorpion flies can detect a male with symmetrical wings either visually or simply by sniffing the chemical signal—the pheromone—he emits. (For some reason, there is an association between the symmetry of a male's wings and his scent, but scientists don't know why.) Given the choice between the pheromone of a male with wings that differ very slightly in length and the cologne of a suitor with matched wings, she will move toward the scent of the even-keeled fly.

Researchers who study elk have determined that the males who possess the largest harems of females not only sport the largest racks of antlers, but also the most symmetrical ones. It turns out that a male elk who loses a fight to another male—and who is thus likely to lose all or part of his harem to that victorious competitor—will grow an asymmetrical segment on his antler the following year, the sorry obverse of a scarlet letter.

By the new evolutionary hypothesis, a symmetrical body demonstrates that the male's central operating systems were all in peak form during important phases of his growth. A well-proportioned body may indicate that the male possesses an immune system capable of resisting infection by parasites, which are known to cause uneven growth of feathers, wings, fur or bone. Or it may signal a more global robustness, one capable of withstanding such threats to proper development as scarcity of food, extreme temperatures or ambient toxins.

In theory, females will select a symmetrical male either for the superior genes that he can donate to her offspring or because he is likely to be in good enough shape to help out with rearing and protecting their young.

"The individuals who have had a good developmental background come out more symmetrical," said Dr. Thornhill. "They're put together better and they'll do better in competition for resources and mates."

The new work is part of the larger study of sexual selection, an intellectually vigorous discipline that is yielding a host of novel proposals of why females opt for one male rather than another. Researchers believe that many outstanding traits found in male animals, from the extravagant plumage of a peacock to the percusive calling of a cricket, have been shaped over generations by female taste, and biologists have sought to understand the sources of that taste. Some scientists have suggested lately that female choice on occasion can be explained by copycat behavior, with females choosing those males whom they have noticed other females fancy.

Dr. Nancy Burley, a professor of ecology and evolutionary biology at the University of California at Riverside, has revealed all sorts of inexplicable preferences among the female birds that she studies. She discovered, for example, that when she put little white feather caps on the heads of male zebra finches, who do not have feather crests of their own, the males became extremely popular with females. By contrast, males given red feather hats gained no advantage from the adornment.

Neither preference appears to have any functional significance, said Dr. Burley. The females could not be choosing white feather caps as evidence of strong zebra finch genes, because zebra finches are not supposed to have hats of any color. Instead, Dr. Burley and others cite results like these as support for the so-called sensory exploitation theory of female choice. By this notion, some female preferences may be the coincidental and still-mysterious reflection of how an animal's sensory system and brain work—what the brain focuses on and what it ignores in the environment—rather than the outcome of a female's careful appraisal of the male's genetic assets.

For every new proposal, there are outspoken detractors, and the study of symmetry is no exception. Critics of the theory complain that some of the differences in body proportion that scientists are now measuring are very tiny, noticeable only when a researcher puts a pair of calipers up to a creature's wingspan. Is it likely, they ask, that these minute variations are obvious to potential mates? They also argue that the case remains to be proved that a symmetrical individual possesses especially hardy genes.

"People are embracing this idea of symmetry because it's something you can go out and measure," said Dr. Marlene Zuk, an evolutionary biologist at the University of California at Riverside. "But it's a black box as to what it means, and what it's indicating to the female."

Whatever its precise relevance to animal sexuality, symmetry is an artistically appealing concept, one that painters, sculptors and architects have been exploring for at least 5,000 years, ever since the Egyptians began building their severely, even rigidly symmetrical temples. In his famed paintings of the School of Athens and the dispute over the sacrament, for example, the Renaissance master Raphael perfectly counterpoised the people on the left side of his canvas with figures of equal numbers and arrangements on the right. "A sense of cosmic order and structure can be conveyed through a symmetrically planned composition," said Kathleen Weil-Garris of New York University, a scholar of Renaissance art. "In both medieval and Renaissance cosmologies, symmetry and proportion, the square within the circle, were part of God's plan."

So, too, does symmetry seem to be part of nature's plan. Most animals have bilaterally symmetrical bodies, with limbs and features mirrored on either side of a central axis, and many flowers are radially symmetrical, all their petals bursting forth in equal arrangements from a central point. Many viruses exhibit an almost mathematical degree of symmetry, as do important structures within the cell that control the cleaving of one cell into two.

Scientists did not begin to appreciate the possible relevance of symmetry to mate assessment until 1990, however. Dr. Anders Moller, an evolutionary biologist at the University of Uppsala in Sweden, was studying barn swallows, a species in which the males have extremely long tail feathers in a wishbone pattern. He had determined that females like long tails on their males, the longer the better. But in attempting to manipulate the feathers experimentally, he made another discovery: females also like symmetrical tails, when one side of the wishbone configuration is the same length and coloration as the other.

Playing around with parameters by adding tail feathers, clipping tail feathers or painting them different patterns, Dr. Moller found that length and symmetry figured more or less equally in female choosiness. In other words, a long, slightly uneven tail and a short, symmetrical tail rated about the same, but when given the chance for a male with a lengthy, balanced tail, there was no contest. Such a well-endowed male was invariably enticing.

In other experiments, Dr. Thornhill and his co-workers upended expectations that the biggest male always prevails by showing that female scorpion flies preferred symmetrical males over bigger males with asymmetrical wings.

The work on symmetry meshed with scientific understanding of how disease and pollution affect animal development. For example, wildlife experts had long known that fish swimming in polluted waters spawn asymmetrically shaped fry. Evolutionary biologists began to suspect that a female animal could make similar assessments of her mate's overall robustness by checking him for symmetry.

Seeking evidence that symmetrical animals are indeed sturdier than their slightly misproportioned counterparts, Dr. Moller has lately determined that barn swallows with symmetrical tails are less likely to be infected with parasites than are males with asymmetrical tails. Through test-tube experiments, he has also discovered that the immune cells of symmetrically appurtenanced males are comparatively stronger, although these results are preliminary and have yet to be published.

In studies with starlings, Dr. Swaddle and his co-workers at Bristol have found that they could influence the symmetry of a bird's plumage by giving it more or less food while it was molting. The less food the bird got, the less proportioned the regrowth of its feathers later on. Hence, the balance of a bird's plumage is a handy and sensitive bellwether of a bird's overall health that season: the more symmetrical the plumage, the more nourished the male, and presumably the better a provider he is for his family, rather as somebody who starts the day with a good breakfast is likely to have more energy by the afternoon.

Symmetry may influence the performance of many traits that a male rallies to lure a female. A cricket creates its chirruping serenades by rubbing its legs together, and scientists are now testing the possibility that females prefer the songs of symmetrically limbed mates over the melodies of a male whose bow and fiddle do not quite match.

The apparent benefits of symmetry are not limited to males. Dr. Thornhill has discovered that the most symmetrical female scorpion flies also are the most adept at gathering and hoarding food, fighting off competitors, dominating their peers and otherwise behaving like members of a ruling caste.

—NATALIE ANGIER, February 1994

Canary Chicks:
Not All Created Equal

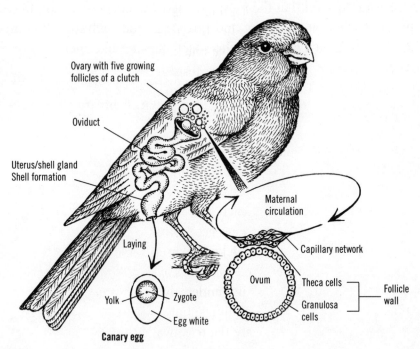

Ovary with five growing
follicles of a clutch

Oviduct

Uterus/shell gland
Shell formation

Maternal
circulation

Capillary network

Laying

Ovum

Theca cells

Follicle
wall

Yolk

Zygote

Granulosa
cells

Egg white

Canary egg

Patricia J. Wynne

HATCHLING CANARIES defy the truism that all baby animals are cute. Born just 14 hours earlier and each weighing hardly more than an aspirin tablet, the three little birds now balled together in a nest are extravagantly repellent, squirming grubs of flesh and fuzz that look less avian than larval. They seem too feeble to be taken seriously, yet simulate the arrival

118

of a parent canary with a puff of air and they start behaving as young birds should: lifting their heads, opening their microbeaks and begging for dinner.

"If they don't do this, if they fail to open their mouths, they will not get fed," said Dr. Hubert Schwabl, a behavioral ecologist at the Rockefeller University Field Research Center in Millbrook, New York. "If they don't beg, their parents will ignore them and they will die."

Each of these birds, however, has been given an extra advantage, tailored to its particular needs. It turns out that its mother has supplied it with a substance that body builders and football players know well as a source of strength, stamina and surliness. Dr. Schwabl has made the unexpected discovery that while the eggs are growing inside her, a mother canary adds to the standard accretion of protein, fat and nutrients a lacing of testosterone, the male hormone. She does this without regard to whether the recipient will be a he-chick or she-chick, but rather to lend the young birds a head start on their development. Testosterone can add mass to a growing body, and it may play a role in the maturation of the spinal cord, allowing the chicks to coordinate their movements, lift their heads and demand their dinner.

Most interesting, Dr. Schwabl has learned that the mother bestows varying amounts of the steroid hormone on her chicks, with the first-laid egg receiving the smallest dose of testosterone, and donations of the potent compound increasing with each successive egg. As a result, the last chick born in a clutch of, say, five eggs gets a dose of testosterone that may be 20 times that given to the first chick. The discrepancy has a dramatic effect. As Dr. Schwabl reported in *The Proceedings of the National Academy of Sciences,* the last-born bird, whatever its gender, always proves to be the most aggressive. It is no bigger than its siblings, but it is the most pugnacious, able to monopolize food and to chase the rest away from its perch.

"This work opens up a new avenue of research for understanding how parents can influence the development of their offspring," Dr. David Winkler, a field researcher at Cornell University in Ithaca, New York, said in a telephone interview. Dr. Winkler wrote a commentary accompanying Dr. Schwabl's paper.

Dr. Schwabl's studies, carried out in a research setting removed both physically and philosophically from Rockefeller's main Manhattan campus, are part of a small but growing effort by biologists to understand the subtle and often surprising influence of hormones on every phase of an animal's development and behavior. In an era when genetics and the glories of DNA reign supreme, and most molecular biologists are fixated on discovering the genes for everything from senility to shyness, researchers of a more naturalist bent are suggesting a different tack.

Genes are only part of the story of any animal's profile, they say, and other influences, like hormones, can contribute to, complicate and in some cases override the innate program inscribed in a creature's genes. And although researchers have long appreciated that a fetus's own steroid hormones, produced by its growing sex organs—testes in a male, ovaries in a female—will in turn help shape the growing animal's body and brain, only recently have they paid significant attention to hormonal contributions from the mother or, in the case of litters, the other siblings in the uterus.

The hormone research exemplifies the scientific philosophy of Dr. Fernando Nottebohm, a prominent behavioral ecologist and neurobiologist who heads the center. Defying the drift toward increasing specialization in science, Dr. Nottebohm seeks to foster a broad perspective in his colleagues, encouraging them to expand their vision of how nature works rather than focus too narrowly on a discrete molecule operating in isolation from the animal that makes it.

"We're trying to integrate everything here, from behavioral work and natural history down to cellular, genetic and molecular biology," he said. "Compared to the immense volume of work going on in neuroscience, there are remarkably few places that pay attention to behavior.

"But behavior is the business of brains; it's what they're designed to do," he added. "If you break everything down into molecules, you end up asking some pretty silly questions. I think looking at biological research as a continuum, from molecular biology up to behavior, offers the greatest wealth for future investigations. There's a beauty in picking up an idea from one end and seeing how it rattles all the way through." That approach has informed his own research, he said, in which he started 25 years ago by asking how birds learn to sing and ended up discovering the regions of the brain in charge of doing so.

For Dr. Schwabl, a capacious view of biology must include an appreciation of other molecules in the body besides DNA, the fashionable object of huge federal efforts like the Human Genome Project. "Genetics dominate other considerations," said Dr. Schwabl. "People forget that the embryo develops in an environment," one that is awash in a ragout of hormones. "Steroid hormones are ancient molecules, they're ubiquitous, they're small and they're sturdy, which means they're not broken down easily," he said. "I see them as the ultimate communication molecules." He points out that pheromones, the molecules responsible for attracting one animal to another, are steroid hormones, as are many defense chemicals in nature designed to tell would-be predators to stay away. "Hormones are the way that one individual communicates with another," he said, "and one generation communicates with the next."

In his recent experiments, Dr. Schwabl discovered that the mother canary donates the testosterone to her eggs early in their development, while they are swelling up from the follicles surrounding her ovaries but before they have been inseminated by a male's sperm. During that rapid egg production, a layer of cells is laid down around the egg yolk able to produce testosterone. Later, during incubation, as the chicks bloom within the shelter of the shells, the testosterone seeps through the yolks and enters their bloodstreams. Dr. Schwabl does not yet know why the mother canary parcels out her testosterone inequitably, but he suspects it is her way of attempting to equalize the birds' initial odds of survival.

Those born first in a clutch have the advantage of hatching a bit earlier and hence getting a jump start on feeding before it is time to fly. "The mother is forced by ecological factors to start incubating her eggs before the brood is complete," said Dr. Winkler. "This creates an asymmetry in hatching times. Testosterone may provide the mechanism for balancing the playing field." It reduces the handicap of the later laid, toughening them up to allow them to make up for lost time by more aggressively demanding their share of food.

Dr. Winkler also suggests that the initial inclusion of testosterone in the eggs makes life easier for the mother, freeing her of the need to pay special attention to the younger chicks.

Whatever the reason, the new work overturns assumptions held by many behavioral ecologists that mothers devote an inordinate amount of their resources to the firstborn, getting around to the younger siblings only in times of plentifulness.

Now Dr. Schwabl is attempting to manipulate the hormone levels artificially, injecting testosterone into newly laid eggs to see whether he can turn the firstborn into a facsimile of the last. The trio of hatchlings is the first fruit of this latest experiment, and it will take about a month to learn whether the manipulations have influenced chick behavior significantly.

He will also be setting up experiments to determine whether birds that receive high doses of testosterone are more successful in ways other than eating. He would like to know, for example, whether they are better able to attract mates or to produce more offspring. Dr. Schwabl warns against the simplistic assumption that heightened aggressiveness is invariably a desirable trait in nature. "Aggression is costly," he said. "Aggressive birds expend a lot of energy, and if they focus only on competing with other birds, they may end up being eaten by predators."

Dr. Schwabl is also examining other bird species to see which ones may use maternal testosterone to influence chick development. Other researchers examining the impact of hormones on development, a specialty known as field endocrinology, have explored the outstanding example of the spotted hyena, in which male hormones from the mother give her female offspring the appearance of having a penis and testicles, as well as the aggressive temperament needed to hold their own against males in the pack, and even dominate them on occasion. Researchers have also studied gerbils, in which the position of a female embryo in the uterus relative to her siblings helps determine whether she is likely to bear sons or daughters. And while some evidence has surfaced recently indicating that human homosexuality may be at least partly genetic in origin, other researchers suspect that the impact of a yet undefined mix of hormones in the uterus is likely to play a role in determining sexual orientation.

Dr. Schwabl, a German scientist with a couple of dashing streaks of white in his brown hair and the sun-creased complexion of a scientist who has spent much of his career observing animals outdoors, works in a facility essentially devoted to putting the life back into the life sciences. The Rockefeller field research center sits on 1,200 acres of rolling wilderness, a mix of second-growth woods, swamps and lakes, home to foxes, badgers, coyotes, an enormous excess of deer, and about 140 species of birds, including great blue herons, common loons, turkey vultures, purple martins, rough-winged swallows, yellow-billed cuckoos, spotted sandpipers and chimney swifts.

The labs are housed in 19th-century stone and wood farm buildings, where the researchers have access to hundreds of canaries and zebra finches, as well as protein purification columns for exacting biochemical measurements and facilities for neurobiological research. Scientists can take insights gained by observing creatures in the wilderness acting on their own terms, and then rigorously test theories in the lab.

—NATALIE ANGIER, January 1994

Rabbits Beware! Some Birds of Prey Hunt in Packs

THE MAJESTIC IMAGE of the lone eagle may often hold true. But scientists are also beginning to piece together a more complex picture of eagles, hawks and falcons as team players whose hunting tactics and cunning intelligence invite comparison with the wolf and the fox.

Eagles, in fact, not only mount concerted and successful attacks on the fox itself; they also deceive monkeys, humans' close relatives, in the deadly game of predator versus prey. By acting together, they are even able to bring down big animals like deer, antelopes and African bushbucks.

Diving, swooping and executing barrel rolls, peregrine falcons double-team rapidly darting swifts, birds that no single falcon could possibly out-maneuver. As the swift veers right and left in a horizontal plane, both male and female come at it from above. The male, smaller and more agile, reverses course once it is below the swift and attacks a second time, from beneath. The multiple assaults drive swifts to such distraction that they fly into obstructions or plunge into water, becoming easy pickings.

And in the Southwest, family groups of Harris's hawks assemble each winter morning, divide into platoons and scour the countryside for rabbits. When one is found, the platoons converge and go on the attack. If necessary, one platoon flushes the prey from brush directly into the talons of the other. If a speedy jackrabbit leads them on a chase, the hawks pursue in relays that keep the quarry running till it drops.

These hawks are "not one whit behind a wolf pack" in their hunting behavior, said Dr. David H. Ellis, an animal behaviorist and raptor expert at the Patuxent Wildlife Research Center of the United States Fish and Wildlife Service at Laurel, Maryland.

Michael Rothman

Team hunting in the air: A group of Harris's hawks poised for a relay attack on a jack rabbit.

As the grimly fascinating evidence accumulates, it is forcing scientists to reassess their longstanding treatment of raptors as solitary predators. Often the birds do hunt alone, and the difficulty of observing them at work has made it hard to discover other kinds of hunting behavior.

But now, according to a study in the journal *BioScience,* there are enough observations to suggest that eagles and their cousins command a wide repertory of predatory actions, including the most sophisticated. This command may be essential to the species' long-term evolutionary survival strategy.

Raptors' newly appreciated prowess reveals "a high degree of intelligence," said Dr. Ellis, the primary author of the paper in *BioScience.* The other authors are Dr. James C. Bednarz, a behavioral ecologist at Boise State University in Idaho; Dr. Dwight G. Smith, a vertebrate ecologist at Southern Connecticut State University, and Dr. Stephen P. Flemming, an ecologist in Sackville, New Brunswick.

Just how bright raptors are relative to the intelligent mammals they kill is unclear and a subject of future research. But in any case, the catalogue of behavior culled by Dr. Ellis and his colleagues from the scientific literature adds up to a chilling picture of raptor craftiness.

Some hunting hawks travel with similar birds, like vultures, to disguise their presence from the prey. A number of raptors follow the leading edges of fires, rising flood waters, moving trains and even people to capture prey flushed by the disturbances. Peregrine falcons have accompanied a moving train for up to six miles for this purpose.

Gyrfalcons in Alaska often followed a trapper to catch ptarmigans, birds that he flushed while tending his traps. In an extreme example, a northern harrier prowled an active bombing range to nab animals and birds scattered by the exploding bombs.

In Venezuela, Dr. Ellis observed a white hawk traveling with a troop of monkeys acting as de facto "beaters," much as humans beat game to the hunters.

Some species, like ospreys, learn of food sources from their flockmates in the colonies where they live, and all the ospreys then go search for prey. And golden eagles in the American West have been known to pounce in semicoordinated attacks on mule deer and antelope, killing them in the winter snow.

But none of this behavior constitutes true cooperative hunting. As used by Dr. Ellis and his colleagues, the term requires that the foraging pair or group be a stable social unit; that some members, in a division of labor, sacrifice their own prospects for a direct kill in deference to the group interest; and that group members share in the spoils.

In the most complex forms, raptors exchange signals to coordinate the hunt and cooperate in hunting outside the breeding season. Many instances suggesting this level of behavior have been observed.

In Manitoba, an adult and a juvenile golden eagle were observed attacking a fox in team fashion: the juvenile, from a height of about 25 yards, dive-bombed the fox from behind, making loud cries to attract the fox's attention.

The fox turned to jump at the juvenile, whereupon the adult, positioned 150 yards aloft, dived silently, striking the quarry in the shoulder blades and knocking it down. The fox, evidently trying to reach cover some distance away, got up and ran again. The tandem strike was repeated. On the fourth attack, the adult eagle sank its talons into the fox and held on. The younger one joined the attack, and after a fearful struggle, the fox was dead.

Does this mean the eagle is smarter than the fox? Not likely, said Dr. Ellis, since "the fox will run the same game on him." In Montana, he said, he has seen foxes distract a golden eagle eating its meal in an attempt to steal the food. If the eagle had simply held its ground, it could have eaten the meal and also killed a fox if it had attacked. Instead, it chased one fox away, creating an opening for the theft. "It's real clear that the fox is smarter," he said.

On the other hand, raptor teamwork appears to signify a higher order of behavior than the cooperative hunting of spiders and ants, in whom it is genetically preprogrammed.

Dr. Ellis has spent months observing the behavior of golden eagles, and he says, "It's hard for me to imagine that they hadn't learned from their mistakes early on and were profiting from that learning, which means they're intelligent rather than practicing something innate." All of this, he says, is grist for further investigation—a daunting task, given the difficulty of studying raptors.

Many other instances of coordinated hunting by pairs of raptors have also been reported. In southern Africa, two black eagles were observed to approach a colony of cape vultures. While the leading eagle flew over the nest and the adult vultures tried to drive it away, the trailing eagle glided in from behind and snatched a vulture chick in each foot.

In a similar ploy, a crowned eagle in South Africa was seen to leave its perch and dive across a valley toward some trees where vervet monkeys lived. It swooped low over the trees and zoomed up and away without pausing. When a monkey climbed out of cover onto a topmost branch to watch the eagle fly off, a second eagle swooped down, grabbing the monkey from behind. Both eagles then flew off together, presumably to feast.

Crowned eagles have also teamed up successfully on the bushbuck, which weighs as much as a small to average adult human and has more than four times the mass of the eagle.

Aplomado falcons in the American Southwest coordinate their efforts this way: the male hunts birds alone in view of the nest, while the female watches without moving as long as the quarry is in the open. But if it flies into a bush, the male hovers overhead and the female flies in, pursuing the prey branch by branch or on foot. When it flies out, the hovering male nabs it.

The most elaborate pack hunting behavior is that of Harris's hawk. Through observation in New Mexico over a period of years, Dr. Bednarz has determined that hawk families—generally two primary breeders, some younger adults and some immature yearlings—form hunting parties each morning. The party breaks up into subunits of one to three hawks. One subunit flies perhaps 250 yards and selects a tree from which to search for prey. The other leapfrogs the first group and takes up another lookout post.

The two groups continually watch each other as the leapfrogs proceed. When one cohort spies a rabbit, its members focus on it intently, sometimes flying toward it directly and emphatically. This is a signal to the other group to join forces for what Dr. Bednarz describes as the "surprise pounce." As many as six hawks dive on a cottontail rabbit, confusing the animal and often killing it on the spot. The family then shares the kill.

If the prey escapes into brush, the hawks use a flush-and-ambush technique. One or two hawks climb into the brush, where they have no hope of capturing the rabbit themselves. But the rabbit panics and runs out to be killed by the other waiting hawks. "This," says Dr. Bednarz, "really can be very effective."

In the relay technique, five or six hawks pursue a fleeing jackrabbit. If the rabbit tries to veer toward cover, the leading hawk dives and deflects it. Losing momentum, this hawk drops to the rear of the pursuing column while a second hawk takes over, and so on. Dr. Bednarz once watched a relay chase cover more than half a mile before the rabbit was exhausted.

Wolves also use both the relay tactic, running the quarry around in a circle, and the flush-and-ambush ploy. And their social structure is almost identical to that of Harris's hawk, leading Dr. Bednarz, who has studied both species, to refer to Harris's hawk families as packs.

Raptors may use coordinated tactics only when solitary hunting does not provide enough food or is too difficult. Harris's hawks, for instance, use them only in winter, when a shortage of smaller quarry like small birds forces them to go after bigger game like jackrabbits, which are several times the size of a hawk.

Dr. Ellis and his colleagues speculate that when raptors hunt alone, they will not even try to capture prey that they know can be captured only through teamwork; it is not worth the expenditure of energy. But bigger game or the quicker capture resulting from group efforts can make the expenditure worthwhile.

—WILLIAM K. STEVENS, January 1993

Meek Bird's Survival Secret:
Picking Protective Neighbors

FOR SAND-COLORED NIGHTHAWKS, choosing the right neighbors can mean the difference between life and death. A biologist working in Peru has found that this species is so helpless against predators that its eggs and chicks can survive only if neighbors protect them.

When predators come to devour precious eggs and nestlings, the mild-mannered nighthawks fly off. But if they have chosen wisely, terns and skimmers from nearby nests dutifully attack the threat. The defenders seem to get nothing in return for their help.

Because so many predators are drawn in by the sight of the nighthawks, the fiercer terns and skimmers, which are often fewer in number, work overtime protecting the nests. As a result, their own nests suffer, with fewer eggs hatching and fewer chicks surviving to fledge.

The sand-colored nighthawk is the first bird known to parasitize the defenses of other nesting birds, according to the study, which was published in the journal *Ecology*.

Nighthawks, which are not hawks at all, are relatives of the whippoor-will and belong to a group with the fanciful name of goatsuckers. Drab and shy, nighthawks typically find safety from their predators by nesting alone in well-hidden places. But along the banks of the Manu River in southeastern Peru, the sand-colored nighthawk nests on the open beach, where it makes an easy target for real hawks ready for a meal.

Drawn by the sight of as many as 200 helpless nighthawks nesting by the riverside, the great black hawk, the black caracara and the roadside hawk drop down to begin feasting on eggs. But interspersed among the nighthawks are birds to be reckoned with: the black skimmer, the large-billed tern and the yellow-billed tern.

"They'll fly at predators, striking them, pulling out feathers, calling very loudly," said Martha Groom, the biologist who studied the birds as a graduate student at the University of Florida. "They make a *Keeee! Keeee!* or barking or yapping sound."

The tough guard species also fight off mammals who eat their eggs, which are attractive to animals like weasels as well as the indigenous peoples living in the Manu Biosphere Reserve. But even a harmless tourist or biologist can be mistaken for a deadly menace.

"I've been hit on the head by a large-billed tern," Ms. Groom said. "It hurts. They draw blood."

She calls the sand-colored nighthawk a parasite because of the harm it inflicts on the species it tricks into defending its nests.

Like other parasites, the nighthawk is extremely dependent on the species that care for it. Terns and skimmers will sometimes abandon their nests if they are disturbed too often by humans or threatening animals. If the guards fly the coop, Ms. Groom said, all the nighthawk nests are raided and destroyed within a few days.

—Carol Kaesuk Yoon, October 1992

The All-Common Crow Has Much in Common with Human Neighbors

CROWS HAVE NEVER been popular.

Farmers, viewing them as a threat to crops, have often greeted them with shotguns. In 1940 the Illinois Department of Conservation killed 328,000 crows with a single blast of dynamite. "Most people dislike crows because they are just like we are," said Dr. Carolee Caffrey, a crow researcher. "They hang around in groups and make a lot of noise. They're troublemakers who like to take the easy way out."

Now, with huge migrations of crows into urban and suburban areas, where they are flourishing on food scraps in trash bins, landfills and garbage pails, more citified folks are getting on the birds' case—with disparaging remarks if not shotguns. But even as some people find new cause to view crows with disgust, these shiny black survivors of urban and suburban sprawl are becoming more popular with scientists.

Dr. Caffrey, for one, a behavioral ecologist at Oklahoma State University in Stillwater, recently completed an eight-year study of western crows living near a Los Angeles golf course. She and others who have been engaged in a flurry of new research are finding that in some ways crows are more like people than people, to paraphrase a popular song. They are faithful to their mates and helpful to their parents, and they maintain a lifelong attachment to their birth families. But they are also wary, wily and opportunistic.

Though scientists cannot document precisely the shifting of crow populations (these commonplace look-alikes are not easy to count), it is evident that in the last several decades millions of crows, once mainly rural, are now breeding and feeding successfully in and around shopping malls, city parks, golf courses and other heavily peopled metropolitan sites.

In most states, crows are protected, having been classified as game birds for which no hunting season is established. But they are still likely to be shot at in some rural areas, though scientists say this can be a grave mistake. As one Midwestern farmer sadly discovered after killing off the crows that he thought were eating his corn crop, that crows have an even heartier appetite for the European corn borer. Without crows to attack this devastating pest, the farmer's crop failed.

Now, shopping malls may be as attractive as corn fields. As many as 50,000 crows may roost near a single shopping center, attracted not only by food but also by the all-night lights that help them spot predators. In turn, scientists now have easy access to crows and are confirming that the birds are playful, resourceful and fast to learn. They are also discovering how communicative crows are, with a complex language that researchers are only beginning to understand.

Perhaps no one has done more to stimulate scientific interest in crows than Dr. Lawrence Kilham, an 87-year-old professor emeritus of microbiology at Dartmouth Medical School whose interest in crows dates to 1918, when as an eight-year-old he took an injured crow to his family's summer home in New Hampshire. There the crow healed and flourished, and all summer long it followed the boy through fields and woods and everywhere else until he had to return to school in Brookline, Massachusetts.

Dr. Kilham, a self-taught ornithologist, studied crows for more than 8,000 hours over a six-year period and produced a seminal work, *The American Crow and the Common Raven* (1989, Texas A & M University Press, College Station), that has helped to elevate the status of crows among both professional and lay bird enthusiasts.

Detailed observations by Dr. Kilham, who had learned to recognize individual crows living on a cattle ranch in Florida without having to mark them, revealed that the birds were cooperative breeders that lived in large family groups. "Teenagers" from last year's brood and even young adults from the year before are likely to hang around the family compound and help out.

Dr. Kevin J. McGowan, a Cornell University ecologist who spends most spring days up in trees that house crows' nests, has discovered that as many as five years of offspring may stay near the parental nest to help out during breeding season. In the last nine years he has marked more than 600 nestlings with colored plastic wing tags in and around Ithaca, New York, an urban and suburban environment.

Dr. McGowan, who says crows are "the least studied game animal in America," has climbed trees to reach nests up to 120 feet high to document this cooperative breeding, which he said is "highly unusual among birds." He has shown that not only are young crows likely to help their parents raise a new brood, but they will also help their brothers raise families.

"My population has a very high percentage of helpers," he said. "Eighty percent of the nests are attended by helpers, and family size around one nest may reach 15."

Crow families like to stay together, Dr. McGowan has found. When older siblings are ready to breed, they often establish territories in or near those of their parents.

Dr. Caffrey, in her study of the golf course crows, found the same cooperative behavior. Older brothers and sisters brought food for the mother and her nestlings, defended the territory and the nest against intruders and predators and stood guard over other family members while they foraged.

In cities, Dr. Caffrey said, a cat on the ground near the nest tree is likely to trigger a loud chorus of alarm calls. If the crows are sufficiently alarmed, they may even mob the cat, as they do predatory owls and hawks that invade their territory. People, too, can be viewed as invaders. Dr. McGowan said he is often mobbed, with 20 or more crows in a family group screaming and diving at him while he tags newborn crows for future identification.

Dr. Caffrey has seen the birds' lighter side, too. "Crows also play all sorts of cool games," she said. She has watched young crows play tug of war with grass and twigs, swing upside down on tree branches like monkeys and play drop-the-stick and fly down to catch it. She even saw one crow roll down a grassy hump on a plastic cup, followed by a sibling who copied the log-rolling sport.

Like human teens and young adults, some young crows leave the family compound for weeks, months or even a year or more at a time. But, Dr. Caffrey said, "from time to time, they come back to visit." One crow she was studying left her family's core area at the golf course at age two to nest with her mate three fourths of a mile away. "But she came back to the golf course every Friday afternoon to hang around with her parents for an hour or so," Dr. Caffrey said.

Dr. Caffrey's crows were unusually laid back, "living and breeding on top of each other and never defending a thing," she said.

"They all fed and roosted together and wandered unharassed into each other's core area," Dr. Caffrey said. "They even let outsiders land in their nest trees. Occasionally two pairs would nest in the same tree, and sometimes they fed their neighbor's kids. In my eight years there I saw no overt aggression, which is unheard of in crows." She speculates that this laissez-faire attitude reflects the abundance of nest sites and food in and around the golf course.

Most crows, though, are highly territorial in their core nest area, where only family members are permitted to alight and feed undisturbed. But on many afternoons, crows that are not nesting are likely to leave the family compound to join others at a huge communal roost, where they feast on worms, insects or grain and—more and more—on human leftovers before settling down together for the night.

Dr. Donald Caccamise, an ornithologist and entomologist at Rutgers University in New Brunswick, New Jersey, said that the combination of cooperative breeding and communal roosting was highly unusual and, on the surface, presented a conflict: "Cooperative breeders need to stay on their territory, but those that roost need to leave their territory. I wondered how crows resolve this conflict."

By fitting crows with radio transmitters that allowed him to track their comings and goings, Dr. Caccamise was able to show that individual crows only occasionally leave home to visit the communal roost, where tens of thousands of crows gather to supplement their territorial meals, in this case by feeding on garbage in the landfills of Staten Island and New Jersey.

"It's just like people who from time to time leave the suburbs to spend the night in Manhattan, where they have a great dinner and a great breakfast and then go home again," he said.

Crows, though, may be gone from their home base for the entire winter, especially when their territorial food sources are covered with snow. Like squirrels, crows store food when it is plentiful, and, researchers say, they seem to remember their hiding places. In one incident in which crows gathered food for future use, a film crew trying to attract hawks released a horde of laboratory mice. Three crows immediately descended and grabbed up 79 of the rodents before any hawk had a chance to cash in on the bonanza.

Crows are also clever thieves. Dr. Kilham watched a crow pull on the tail of a river otter that was holding a fish in its mouth. When the agitated otter dropped the fish, the crow's compatriots swooped down and grabbed it. Crows have also been seen pulling up the unattended lines of ice fishermen and stealing either the bait or the catch.

The antics of crows are often easy for amateurs to observe. But even unseen, their language tells much about their behavior.

Dr. Cyndy Sims Parr, who just received her doctorate at the University of Michigan based on a study of crow vocalizations, said the birds' long-distance "caws" have different meanings determined by their form and rhythm and how they are strung together. Various vocalizations can mean, "Stay out of my territory," "Watch out, someone's after your lunch" or "Help me get this predator out of here." For example, she said, a string of *ko-ko ko-ko ko-ko* means, "Neighbor, you're trespassing."

"The long-distance vocalizations, of which there are 15 or 20, function like a standard bird song," Dr. Parr said. In addition, there are the many soft sounds crows make when talking with family members, including rattles, growls, gargles, coos, squawks, squeals and plaintive *oo-oo*'s. For example, young crows squeal when teasing one another, and the *oo-oo*'s are begging sounds uttered by hungry chicks and females on the nest.

Dr. Caffrey has noted that Eastern and Western crows have different accents. There is a crow on just about every movie sound track, Dr. Parr said, and Dr. Caffrey said she can tell whether the movie was shot in the East or the West by how the crow sounds.

—JANE E. BRODY, May 27, 1997

Mick Ellison

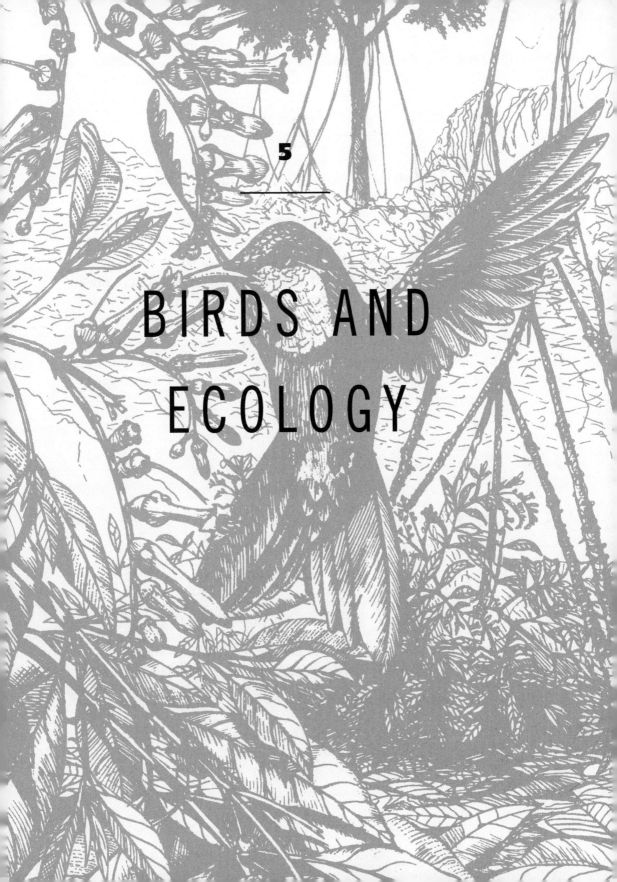

5

BIRDS AND ECOLOGY

With their cheerful songs and carefree flight, birds may seem like an airy adornment on nature's cake.

As the following articles show, the case is often quite the opposite. Whole ecologies may depend on a critical species of bird, like the forests of Sulawesi in Indonesia where many trees require the dispersal of their seeds by the red-knobbed hornbill.

A recent study has shown how critically American forests depend on songbirds to keep down the populations of voracious insect predators.

Birds play central roles in the interdependent communities of plant and animal, whether in pollination, seed dispersal or restraint on insects.

For Clues to Antarctica's Health, Ask a Penguin

AS TWO GENTLE HANDS raised her from her nest, a nine-pound Adélie penguin cocked an eye, sized up the intruder and jabbed her sharp beak at an exposed wrist. Without stopping to dab at the blood from his wound, Brent R. Houston quickly slipped a numbered aluminum band around one of the bird's flippers and restored her to the pair of eggs in her pebble nest.

"They're not exactly the cute little critters most people think," said Mr. Houston, a wildlife biologist from Old Dominion University in Norfolk, Virginia. "They lead hard lives, they're survivors and they may have something important to tell us about the state of the world. You can't help but respect them."

As the breeding season for Adélie penguins in the Antarctic Peninsula draws to a close, these quintessential symbols of Antarctic wildlife have become the objects of one of the most comprehensive biological studies ever undertaken at the bottom of the world.

Penguins, like people, are near the apex of the food chain, and some biologists compare penguins to the canaries miners once used to warn them of the presence of deadly gas; if penguins begin to fare badly, the ecosystem as a whole, human beings included, may also face trouble. With mounting evidence that synthetic chemicals have opened a dangerous breach in the atmosphere's ozone shield, marine food shortages created by the pernicious effects of solar ultraviolet radiation leaking through the shield might eventually affect penguins—and perhaps even human beings.

The waddling gait, tuxedo-like markings and funny mannerisms of *Pygoscelis adeliae,* named by the 19th-century French explorer Dumont d'Urville after his wife Adélie, have long caught the attention of cartoonists. The charming birds have also attracted human predators. The body fat of

139

Adélies has served as lamp oil, and dried penguin flesh shredded into "pemmican" field rations nourished many an explorer.

But although Adélie penguins have been observed and studied for nearly two centuries, quantitative data are still lacking as to precisely how these birds balance the energy they use swimming, hunting, breathing, staying warm and reproducing with the energy they must obtain from food. The balance is believed to be sensitive and precise, and Adélies are not considered capable of adapting their behavior to major changes in their habitats. Thus, for example, a major change in the availability of krill—the shrimp-like staple of the Adélie diet—could mean the difference between survival and starvation. Krill, in turn, are believed to be somewhat vulnerable to ultraviolet rays, as are the algae on which they feed.

But even without major changes in the overall conditions of their habitat, penguins do well in some years and badly in others; such fluctuations are called "normal variability." So scientists looking for environmental trends must know what is truly "abnormal" for a penguin colony, and what therefore indicates a real trend. Biologists must continuously log normal variations in population, breeding success and other conditions over an extended period, something that has never been done before.

This year marks the beginning of a National Science Foundation project called the Long Term Environmental Research Program in Antarctica, a six-year investigation during which many Antarctic creatures, including penguins, will be intensively and continuously observed.

Part of the project entails banding more than 150 breeding pairs of penguins so that their lives can be more or less tracked from year to year. One of the centers for this study is Torgersen Island near the United States Palmer Station—a small island where about 8,000 breeding pairs of Adélies gather for a few weeks each year, making the rocks slippery with guano and filling the air with their gabbling chatter.

After banding a representative sample of penguins and other Antarctic sea birds, scientists from Old Dominion University stand watch over the rookeries, logging the comings and goings of band numbers, using binoculars to avoid disturbing the birds. When a penguin fails to return to its nest after a 10-day foraging trip at sea, it is presumed to have fallen prey to a marauding leopard seal or some other predator.

"They don't seem to be quite as monogamous as was generally believed," Mr. Houston said. "They nearly always return to the nest they share with some mate, but the mate seems to be less important than the nest, and there's a certain amount of mate swapping and replacement."

Penguins build nests with the only movable material available to them: pebbles. To scrape together a concave nest about 10 inches in diameter, a pair of birds may have to forage far and wide to find enough pebbles, each of which must be about an inch in diameter. In a penguin rookery every available pebble is incorporated into a nest, and some biologists believe that one of the limits of penguin population growth is the availability of pebbles.

Real estate values in penguin colonies vary widely, the most desirable spots being at the center of a large community, where nests are less accessible to predators. Competition for nesting sites is fierce, and latecomers to a colony must usually settle for exposed, outlying sites. It appears that in these bloody competitions, the male penguins fight for nesting sites while the females compete for males, at which time some "divorces" usually occur.

Complicating penguin tallies, each colony always has a number of "loose penguins," mateless birds who waddle around aimlessly with pebbles in their beaks, searching for a mate willing to accept the pebble and start a nest and family. Young penguins generally remain "loose" for their first three years of life before settling into more or less permanent domestic arrangements.

It is not easy to tell penguins apart, although one inhabitant of this rookery who has returned every year since 1985, "Blondie," stands out because she is an albino. Even the sex of a penguin is not obvious, although males have slightly larger heads than females. "The surest test," Mr. Houston said, "is to look for mud on a penguin's back. When it's there, you know you're looking at a female that's just mated with a muddy male."

For the first time, miniature recorders are being attached to Adélies to register their swimming and hunting strategies. Dr. Mark A. Chappell, a physiological ecologist from the University of California at Riverside, glues a strip of Velcro tape to the feathers on the back of a penguin to which he attaches a sensing and recording device that weighs only two ounces.

The battery-operated device, designed and built by the Wildlife Computers Company of Woodinville, Washington, records the duration and depth of each dive a penguin makes on one of its foraging expeditions.

When the bird returns to its nest to regurgitate food for its chick and to relieve its mate, the recorder is removed and its data are transferred to a computer. The Velcro tape comes off when the penguin molts.

Sometimes birds equipped with recorders fail to return—a loss not only of data but of a device that costs $1,400. "Despite the occasional losses, we're learning many things," Dr. Chappell said. "For instance, we now know that in a typical foraging sequence, a penguin makes 70 or 80 dives, each one lasting a minute or two and reaching a depth of something less than 100 meters. We're seeing indications that penguins may be capable of locating krill swarms by sounds the krill emit. The bird catches between 100 and 1,000 krill with each dive. A full penguin contains about 900 grams of food, more than a quarter of its body weight." A single mature krill is a little over an inch long, and biologists who have tasted them describe the flavor as like that of "fishy cellophane."

Such facts are essential in computing the energy budget of a penguin, an indicator of how sensitive the birds are to changes in their habitat.

To measure penguins' metabolism, fat content, food consumption and other factors, blood samples are taken and some stomach contents are extracted by a messy pumping process called "lavage," to which penguins strenuously object.

"Popular opinion notwithstanding," Dr. Chappell said, "these are not friendly birds. They're strong, tough and aggressive, qualities essential to survival in such a hostile environment." Despite the rigors of Antarctic life, he said, a lucky Adélie can reach the age of 14.

Brown skuas, large gull-like predators, haunt Adélie rookeries, constantly trying to grab penguin eggs or kill unguarded chicks. But the penguins sometimes fight back.

"The other day we saw the tables turned on a skua," Dr. Chappell said. "A skua grabbed an egg and had actually taken off when the Adélie parent lunged and caught the skua by the leg and pulled it down. At that point several other penguins joined in the fray and beat up the intruder pretty badly. The skua limped away looking very puzzled."

Investigators prefer studying penguins to studying many other birds because they are accessible.

Dr. Chappell finds them notably easier to investigate than another of his favorite subjects, the Colorado hummingbird. "We like to work on penguins," Dr. Chappell said, "because they're convenient. They're not tame, but they're not very frightened of people, either. As a penguin ages it seems to become rather senile, although they're not too bright to start with. A penguin can be fooled by putting a wooden egg in its nest, while a skua that tries to steal the egg quickly recognizes the deception and abandons the fake."

The time-and-depth recording device has given scientists some surprises during the last few weeks. "Adélies are not as hardworking as I had thought," Dr. Chappell said. "After they leave their nests to go out foraging for their chicks, they spend a lot of time just resting, not diving. We've learned that Adélie parents know exactly how much food it takes to rear a chick for the seven weeks before they abandon it, and they catch the bare minimum of food, no more."

Adélies and cormorants in this region are facing a new threat to their environment that comes from nature rather than human beings, and scientists are at a loss to explain it. For some reason, the population of elephant seals has been booming in recent years, and although these huge animals do not eat birds, they love to lie on the smooth rocks used by birds as rookeries, crushing eggs and nests. During the breeding season Adélies rarely travel more than 30 miles from their nests, and the growing crowd of seals threatens to ruin many of the nesting sites available within this radius.

But for richer or poorer, the domestic lives of Adélie penguins is going to be better documented than ever before. "We're at last getting the equivalent of a video record of life in Antarctica rather than a mere collection of snapshots," Dr. Chappell said, "and it is revealing to us a new and vital dimension of this fascinating habitat."

—MALCOLM W. BROWNE, December 1991

More Than Decoration, Songbirds Are Essential to Forests' Health

THE MIGRATING SONGBIRDS that brighten North America's forests each spring with dazzling colors and melodious strains have been regarded by many scientists as mere decorative frills that play a limited role, if any, in the functioning of their ecosystems. But a new study provides what researchers say is the first evidence that these birds, some of which are in decline, play a crucial role in maintaining the health and productivity of forest trees.

In the study, published in the journal *Ecology*, two researchers have shown that these voracious predators can protect trees by eating the bugs that chew the leaves. As a result, scientists are predicting that the decline of these birds could eventually threaten the productivity of forests in the United States as well as in the birds' wintering grounds in the tropical forests of Central and South America.

"It's really a spectacular result," said Dr. Donald R. Strong, an ecologist at the University of California at Davis, "and it's got to be generally true. The conservation importance and forest management importance is profound. We see that in order to protect our forests we've got to have these birds, some of which are becoming quite rare."

Dr. Richard T. Holmes, a bird ecologist at Dartmouth College in Hanover, New Hampshire, said: "It's a very exciting paper. It's the first time that there have been data to show that birds feeding on insects have an impact on trees."

Working in the forest at the Tyson Research Center, a reserve near Eureka, Missouri, the researchers enclosed white oak trees in netting cages. The cages were built with holes big enough to admit all the insects that might attack the trees but small enough to keep out the birds that would eat the insects.

For two years biologists monitored the insects flying in and out of the cages. Most were moths and butterflies laying eggs that would hatch into leaf-devouring caterpillars, the major insect enemies of these economically important timber trees.

What they found was that the plants that were caged to keep out birds were plagued by twice as many insects and had twice as much of their leafy green foliage eaten as uncaged plants. As a result, the damaged plants produced fewer leaves in following years than those trees cleaned of their pests by birds.

"There's been a lot of argument and controversy about the role that birds have in ecosystems," said Dr. Robert J. Marquis, a plant ecologist at the University of Missouri in St. Louis, who was an author of the study. "Some say they do no more than provide weekend activity for bird watchers. We've shown that the birds can have an effect."

Conservationists welcome the work as the first piece of evidence that the colorful dickeybirds of spring do have a significant ecological and economic importance.

Dr. Tom Hinckley, Bloedel Professor of Forestry at the University of Washington in Seattle, said: "These birds in a sense enrich our environment, whether it's urban or rural, but it's sometimes difficult to put a value on them other than enrichment. Now here's a case where their value can perhaps be translated to a clear economic value."

Dr. Christopher J. Whelan, who holds the Emily Rodgers Davis Chair in Ecology at the Morton Arboretum in Lisle, Illinois, and was the other author of the paper, said the migrants made up more than two thirds of the nesting bird species in the forests.

During spring migration at the study site in the Ozarks, 40 to 50 bird species arrive to eat the insects on the fresh white oak leaves. Dr. Whelan said some 30 of those species stay on to nest, including red-eyed vireos, American robins and scarlet tanagers. All appear to help the white oak.

Though many researchers agree that several species of these migrant songbirds are decreasing, the reasons remain controversial. Researchers have attributed the decline to forest destruction and other disruptive forces here and in their wintering grounds.

In addition to documenting beneficial effects for plants, researchers say the new study may provide clues to the birds' decline. Dr. Scott Robinson, an ornithologist with the Illinois Natural History Survey in Champaign, said that the migrant birds' ability to depress insect numbers so greatly on white oak suggests that competition for insect prey on breeding grounds in North America may be important to the birds' survival and reproduction, as well as a factor limiting their numbers.

Based on the new study, ecologists predict trouble ahead for forests, but they caution that for the moment such predictions are speculative. Whether any trees besides white oak will actually suffer remains to be seen.

Researchers say that even if there is an important effect on trees, the damage due to loss of birds may be difficult or impossible to tease out from that due to other causes. Although they cannot point to specific damage resulting from the loss of birds, some researchers say they fear damage to trees may be under way.

"My feeling is that it's going on as we speak," Dr. Whelan said. "In some of the small forest fragments that have very few birds in them, this may very well be happening right now."

Ecologists have known for some time that trickle-down effects along a food chain can be very important in aquatic systems. In these so-called trophic cascades, top-level predators, like fish that eat smaller fish, radically increase or decrease the numbers of creatures at the bottom of the cascade by eating more or fewer of those that act as predators further down the line. But scientists had remained skeptical about whether the terrestrial equivalent of these cascades—involving birds, insects and plants—were of much importance. Birds, it was supposed, would not be powerful enough predators to have an effect all the way down to the plants.

"The really exciting thing about this paper is, now we can believe that the birds are playing a big role, having a hand in this," said Dr. Strong.

The authors even propose that the birds constitute an ecologically correct form of pest management.

"Some people say insecticides are the way to go," said Dr. Whelan, "but, actually, what we show is that birds were as effective as insecticide. So if you have the birds, you don't need insecticide." Many ecologists say that anyone who is interested in reducing pests in forest tracts and increasing forest productivity should protect the beneficial birds.

Many options for increasing bird numbers are already well known. For example, many of the birds declining most quickly prefer forest interiors.

One way to protect them is simply to prevent the fragmentation of forest tracts. Researchers say such fragmentation not only decreases the stretches of protected interior forest but provides habitat for species that like edges, like raccoons and blue jays, which often prey on the eggs and young of the declining bird species.

The researchers will continue their work by examining the effects of these migrant insect-eaters on tropical forest trees in their wintering grounds in Costa Rica.

—CAROL KAESUK YOON, November 1994

Hornbills Are Lynchpins of Indonesian Rain Forest

AN ECOLOGIST studying a spectacular fruit-eating bird, the red-knobbed hornbill, in Indonesia has found that it seems to play an important role in maintaining and regenerating the rain forest.

The birds have their work cut out for them: the island of Sulawesi, where the study was led by Dr. Margaret Kinnaird of the Wildlife Conservation Society, has lost 90 percent of its richest lowland forest to agriculture. And the hornbills depend on the forest, even as they help it. If Sulawesi, also known as Celebes, continues to lose forest at its current rate, Dr. Kinnaird warned, the red-knobbed hornbill's future is uncertain.

"Hornbills are the farmers of the Sulawesi forest," Dr. Kinnaird said. "They travel long distances in search of ripe fruit from more than 50 different trees, particularly energy-rich figs, and they deposit the seeds not only in primary forest but in secondary forest and burned habitats."

Tiny fig seeds are defecated undamaged and sprout in crooks and branches high in the forest canopy, Dr. Kinnaird reported in *Natural History* magazine. "Larger seeds of fruits such as nutmegs and wild mangoes are spit out, often more than an hour after the fruit is harvested," she wrote. "By that time a hornbill can have traveled a mile or two from the parent tree and deposited the seed in a spot where it can more likely germinate free of competition."

The relationship of hornbills and fruits is reciprocal, Dr. Kinnaird said. "The redknobs are critical agents of seed dispersal, cultivating their own lush garden," she said.

The red-knobbed hornbill, *Aceros cassidix,* is endemic to Sulawesi and adjacent islands and is among the largest of a group of some 53 hornbill species that are found from the African savannas to the tropical forests of the Philippines. Thirty inches long and weighing more than five pounds, the bird carries a huge curved bill that is topped, in the male, by a large red casque.

The female has a somewhat smaller bill and a smaller, yellow casque. The casque consists of horny skin covering an air cavity that apparently serves as resonating chamber for hornbill honks and barks that can be heard 300 yards away. Moreover, air passing through a flying hornbill's wing feathers produces a loud, jetlike *swoosh* that can also be heard at a considerable distance.

In addition to the casque, the ivory-colored bills are decorated at the base with red chevrons, and the number of stripes may signal the bird's age. "According to local lore, a new stripe is acquired each year," Dr. Kinnaird said.

Most hornbills nest in tree cavities, and they are noted for the female's habit of sealing the entrance with mud or, in the case of the redknob, excrement held together by fig seeds.

"The local volcanic soil never becomes muddy even during the rainiest months," Dr. Kinnaird said.

Nest-sealing by female hornbills is believed to be a defense against predators, and the incarcerated parent and chicks receive food deliveries from the male through a narrow slit while fecal matter is thrown or squirted out of the opening.

The Sulawesi bird's nesting period from sealing to fledging lasts 133 days, but after about 90 days the female breaks out of her prison and forages on her own. But she offers little help in feeding the single offspring and even steals fruit being passed to the chick.

The red-knobbed hornbill has a fleshy, electric-blue throat pouch that can be stuffed with a pound or more of figs for airlifting to its nest. "The pouch doesn't stretch as much as a pelican's," Dr. Kinnaird said, "but you can see the lumps and tell when the bird is coming back with quite a load. I watched one male cough up 265 figs of a single species. You can identify every item that the bird delivers, which is one of the reasons I couldn't resist studying it. Not much work has been done on hornbills, and they're a very sexy species."

Dr. Kinnaird's research was conducted at the 22,000-acre Tangkoko-Dua Sudara Nature Reserve on Sulawesi's northernmost tip. The crab-shaped island, east of Borneo, consists of four mountainous and finger-like peninsulas. She was assisted by her husband, Dr. Timothy O'Brien, also an ecologist, and graduate students from the local university. A report describing how high numbers of red-knobbed hornbills are linked to fig production on the forested slopes of Tangkoko volcano appeared in a recent issue of *The Auk,* the journal of the American Ornithologists' Union.

Tangkoko rises from the shore of the Sulawesi Sea to a height of 3,400 feet, and Dr. Kinnaird describes the mountain as a biologist's dream. "Unlike most tropical rain forests, the understory is so open that you can appreciate the size of the trees with their enormous buttresses," she said. "And you don't have to fight your way through thickets and be clawed by everything thorny.

"Because of the volcanic sand," Dr. Kinnaird continued, "I wasn't covered with leeches every day. And Sulawesi's climate has two seasons. The hornbills begin breeding in late June during the hot, dry season, so I didn't have fungus growing out of my hair and on my skin like scientists who work in Borneo. Another blissful thing about this forest is a palm called the woka, with huge fan-shaped and water-repellent leaves. If it starts to rain you sit under a woka leaf or two, and if you look around there'll probably be a monkey nearby doing the same thing."

The monkey is the crested black macaque, once known as the Celebes ape. "They have only a nubbin of a tail and spend most of their time on the ground," Dr. Kinnaird said.

A biogeographical boundary drawn by the 19th-century naturalist Alfred Russel Wallace runs between Borneo and Sulawesi, she noted. "Typical Asian plants and animals are found west of Wallace's Line," she said, "but on this side you start picking up these strange endemic species."

The red-knobbed hornbill is one of only four hornbill species that have crossed Wallace's Line. But its numbers in the Tangkoko forest are four times greater than the combined density of all seven hornbill species on neighboring Borneo, Dr. Kinnaird reported. The volcano's slopes support as many as 200 hornbills, both resident and transient birds, per square mile during peak fig fruiting periods. And in one breeding season, the scientists counted 60 active nests in a 2.3-square-mile area.

Redknobs, Dr. Kinnaird said, fly as far as 13 miles a day to feed on the fruit of more than 30 species of figs that abound year-round on Tangkoko's slopes at densities of up to 11 trees per acre. "Clusters of figs droop from branches or protrude like cauliflower heads directly from trunks," she wrote in *Natural History*. "Some are the size of peas; others as big as plums. Colors range from blood red and day-glow orange to lime green. The thick, sweet aroma of rotting figs permeates the air." Fig crops on a single large strangling fig ranged from 100,000 to one million fruits.

Nonfig fruits tend to occur on Tangkoko in small crops and mature slowly, Dr. Kinnaird said. "Only a few ripe fruit are available on a given day and they are rapidly eaten by hornbills, imperial pigeons and macaques."

In their paper for *The Auk*, the scientists propose that the abundance of figs, combined with abundant tree cavities, explains "the extraordinary red-knobbed hornbill breeding density" in the Tangkoko reserve. Figs account for 70 percent of the bird's diet during the breeding season, their studies show. And an immense tree called the nantu, which is the hornbills' favorite nest site because it is prone to heart rot, is common in the forest.

The density of fig trees and fruit production at Tangkoko, they wrote, are exceptionally high compared with other Southeast Asian forests, where hornbills usually are found in relatively small numbers. Moreover, while figs and red-knobbed hornbills are also found in secondary and burned forests or regenerating agricultural lands, the birds prefer primary habitat where larger fig trees bear larger crops.

—LES LINE, January 1996

Hummingbirds Bear Mixed Gift to Flowers in Tropical Forest

EVERYWHERE throughout this jungle the pouting, blood-red lips of *Hamelia* flowers draw in hummingbirds, which stop for a moment to sip nectar and shed a dusting of precious pollen. But like many of the rain forest's complex relationships, this trade between bird and flower is more intricate than it looks. Researchers have found that in addition to pollen, these plants receive flower-attacking mites that come cruising through the forest riding on the beaks of the dazzling birds.

The life-teeming tropics abound with species that have grown to depend on one another in intimate ways. The mites are at the heart of one of the most complex of these interrelationships. Too small to travel far by walking, the tiny relatives of ticks and spiders have learned how to exploit the flower-to-flower public transportation system provided by nature. With precise timing, the mites must leap onto the blindingly fast hummingbirds as they drop in at a flower. Then, when the hummingbird next visits the same species of flower, the mite must get off by tearing down the beak, running as fast as a cheetah in terms of its body length, if it is to alight at its next Eden before the bird leaves.

Two studies in the journal *Biotropica* have put a new twist on the bizarre tale of these consummate miniature travelers whose lives revolve around ephemeral flower homes and the unpredictable comings and goings of their winged transport. It is a tale constructed from over two decades of research led by Dr. Robert K. Colwell of the University of Connecticut in Storrs.

"It's really a fascinating story," said Dr. Shahid Naeem, a community ecologist at the University of Minnesota in St. Paul. "These little, tiny creatures can't hope to come across another plant by walking, say, from a bromeliad 150 feet up in a tree. But you can do it on a hummingbird going

from flower to flower. It really is quite a neat example of evolution and ecology with so many different forces at work."

Researchers say that the long, careful analysis of this system has revealed just what it takes to make it in the rain forest for an animal a little larger than the dot on this "i." Dr. Brian Farrell, an evolutionary biologist at the University of Colorado in Boulder, said the system had yielded answers on many fronts, "because of Rob's great and painstaking knowledge."

Michael Rothman

"Knowing this kind of system in detail takes 20 years," he said. "Rarely do you see this kind of dedication, and that's what's required, especially in the tropics."

It is midnight in the forest and a restless crowd is gathering. Poised at the base of imminent blooms, the thousands of mites on this slight *Hamelia* are waiting for the plants' clock to strike 1 A.M., when the flowers' petal doors will open and the mites will go rushing in. Once inside they will begin their small hours' feast, eating pollen and lapping up a dawn burst of nectar, all the while indulging in an orgy of mating as males battle to control harems of females.

Most of the mites will stay on this shrub all their lives. But an adventurous few—females gambling on hitting the jackpot of an uninhabited cluster of flowers, and males risking all in the hopes of finding a spate of virgin mates—will seek to explore the terra incognita of the vast rain forest. So these specks of animals, unable to depend upon their own eight negligible feet, take passage on hummingbirds to make their escape.

Dr. Colwell said it was in 1969, while teaching a tropical biology course in Costa Rica, that he noticed that the mites were only to be found in the bright, red flowers visited by hummingbirds.

"The mites themselves were already known five years before," said Dr. Colwell in an interview at the La Selva Biological Station in Costa Rica, a research station of the Organization for Tropical Studies and one where he has done much of his research. "But no one had actually looked at them or knew anything about what they were doing."

Pointing to a rufous-tailed hummingbird stopping for a few seconds at flower after flower, he noted the dilemma of the traveling mite. Each species of mite is extremely particular about which species of flowers it will use. "That hummingbird could have three species of hummingbird flower mites in its nose," he said. "We've found as many as five. These mites have only two or three seconds in which to make a decision about disembarking from a bird and then executing—that's not a very long time. But they hardly ever get off on the wrong plant. Only about 1 in 200 makes a mistake."

How do these blind wayfarers, operating with the barest makings of a head and brain, discern different species of flowers and commit themselves to alight in a mad dash—12 body lengths per second—down their transport's beak?

What researchers have found is that the mites recognize the plant species they prefer by the bouquet of their nectar. In tests in which mites were offered one nectar versus another, they consistently went for the familiar honeys of their preferred species. Holed up in the hummingbirds' nostrils and awash in each of its breaths, the mites quickly perceive when they have reached the right stop.

There is good reason for their careful choices. For some mite species, like *colwelli,* which another researcher named for Dr. Colwell, getting off on the wrong flower can mean death, since some of the flowers they mistakenly visit harbor other mites that will kill the visitor. The mite also risks being stranded without hope of finding a mate.

Researchers say the mites' habits have provided ecologists with a novel answer to the question of why plant-eating insects and mites are so often restricted to attacking only one or a few species. Dr. Farrell explained that typically researchers have focused on the importance of plants' defensive chemicals in keeping insects and other pests from using just any species they come across.

"One of the important things people want to know is what governs that kind of specificity for a host plant in general," he said. "In this case it's optimizing mate encounter, and that's a pretty neat discovery."

Once on board, the mites must endure whatever travels the bird undertakes. Biologists established that some hummingbird flower mites found seasonally in California in Indian paintbrush plants apparently travel more than 1,000 miles, held captive in the noses of migrating hummingbirds flying north from the mountains of Mexico.

But there are dangers for the homebodies who live out their lives on one plant as well as for the more adventurous ones. Researchers say that the mites face the constant problem of getting out of their flowers in a timely fashion, especially the many blooms that last but a single day before they are unceremoniously dropped from a plant.

Yet in spite of all the risks of their strange lifestyle, the mites are immensely successful. They have been detected in hundreds of plant species and on more than 60 species of hummingbirds, from California through Central and South America. And until recently these ubiquitous little blebs of life were thought to be relatively harmless.

Now in a new paper by Dr. Colwell and in another by undergraduates from Carleton College in Northfield, Minnesota, a different light has been cast on the nature of these stowaways. The mites are so voracious that they can eat nearly half of the precious nectar that would otherwise be available to the birds, as well as one third of the pollen.

"Taking that much pollen is nasty," Dr. Colwell said. "That's pure gametes," or reproductive cells. "They are also a huge problem as a serious competitor with hummingbirds. The mites are taking equivalent proportions of nectar."

So, researchers say, the mites are not only taking free rides, they are robbing food from those who carry them.

"It's interesting because there are interactions below the surface that we have sometimes overlooked," said Dr. Raymond Heithaus, editor of *Biotropica* and an ecologist at Kenyon College in Gambier, Ohio. "Lots of people have studied the interaction between hummingbirds and plants but have not included this extra effect that mites or other consumers of nectar might have on that relationship. It forces us to think more carefully about this."

Dr. Colwell remains enthusiastic about the mites he has studied for over a quarter of a century. "I love them," he said. "I think they're terrific."

—CAROL KAESUK YOON, May 1995

6

MIGRATION

Migration is one of the most fascinating of birds' many capabilities and is still far from being completely understood.

The earth's magnetic field has long been recognized as one cue to navigation, but many species seem to use other cues as well, from the direction of polarized light to the positions of certain constellations.

Migration is an ancient behavior that each species has developed over the course of evolution. In many species the migration route is genetically programmed. So biologists were surprised recently to find a European wood warbler known as the blackcap develop a new migration route before their eyes.

In larger birds like geese and cranes, the young generation must learn the migration route from experienced adults. While some biologists are trying to tease apart the secrets of avian navigation, others are teaching such birds how to migrate.

Migrating Birds Set Compasses
by Sunlight and Stars

Nocturnal Navigators
Scientists know that birds use constellations, the Big Dipper, for example, as navigation guides when they fly long distances.

Birds' Internal Magnetic Compass
Another navigational aid for migrating birds is an internal compass sensitive to fluctuations in and the direction of the earth's magnetic field. Many species of animals and even bacteria exhibit this sense.

Magnetic North Pole

Magnetic Equator

Geographic Equator

N

S

60,000 nT

60,000 nT

Magnetic intensity is not uniform at all latitudes.

Dimitry Schidlovsky

SCIENTISTS have known for more than two decades that birds and many other animals navigate with help from the earth's magnetic field. It now appears, however, that birds must frequently calibrate their sense of magnetic direction using many nonmagnetic cues, including the natural polarization of daylight.

New research, moreover, suggests that in at least one species, the bird's eye detects the earth's magnetic field using the energy of daylight to sensitize a chemical in the retina to the earth's magnetic polarity.

The navigating skills of birds, amphibians, reptiles, fish and even mammals seem to depend on complex arrays of sensory cues, including magnetic fields, visual patterns, sounds and even smells, interacting in subtle ways. Some scientists say the latest investigations support their contention that magnetic fields have biological effects on human beings as well, a suggestion that has been discounted by many physicists and physiologists.

In any case, two papers in the British journal *Nature* imply that biological navigation systems are even more complex and subtle than many investigators had believed.

In the first paper, Dr. Kenneth P. Able and his wife, Mary A. Able, both of the State University of New York at Albany, presented experimental evidence that savannah sparrows, which migrate between the Northeast and the Deep South or Mexico, not only see patterns of polarization in the daylight sky, but use the orientation of these patterns as a navigation aid for calibrating their magnetic directional sense. Surprisingly, the birds are not influenced by the position of the sun itself, but by the directional polarization of sunlight scattered by the atmosphere—the "Rayleigh scattering" responsible for the blue color of the sky.

"Probably, these birds see a very dark polarized band in the sky 90 degrees from the sun," said Dr. James L. Gould, a biologist at Princeton University. "The tilt of this band, which is invisible to human eyes, would tell a bird the position of the sun and his orientation on earth." The polarization would be visible to a bird only under a clear sky; an overcast would block it completely.

Many fish, including tuna and salmon, are also excellent navigators, but to use polarized light as an aid to calibrating their magnetic sense they would have to swim very close to the surface, because polarization is filtered out by water at depths greater than a few inches, Dr. Able said.

Although most human beings cannot see patterns of polarization in light with the naked eye, they can see them, including the polarized band seen by the savannah sparrows, by holding up polarizing sunglasses or camera filters and observing the sky through them. "We have no idea at all how birds see polarization patterns in the scattered light of the blue sky," Dr. Able said, "but they do seem to sense it somehow."

The Ables conducted their experiments by keeping the savannah sparrows in cages exposed to the sky. When the time came for the birds to migrate south, their restless movements were recorded on the paper floors of their cages by their ink-smeared feet. The birds faced the direction in which they would migrate if they were free.

But in some of the tests, a large electromagnet surrounding a test cage was turned on to shift the magnetic field felt by the birds. The field was shifted 90 degrees, so that the artificial magnetic north imposed by the magnet would cause a compass needle in the cage to point to the west. In another variation, a transparent plastic sheet was placed over the cage to remove polarization from the daylight reaching the birds.

The Ables found that under the plastic depolarizing filter, a shift in the artificial magnetic field around the cage caused the birds to shift their preferred direction correspondingly; they ignored the visible sky as a navigational guide and relied on the spurious magnetic cue. But when the depolarizing filter was removed so that the birds could see unfiltered daylight, they ignored the misleading magnetic field and oriented themselves instead toward true south, as indicated by the naturally polarized daylight.

"It appears that the magnetic navigational sense in migrating species is used as a backup when celestial cues are lacking," Dr. Able said. "When the birds are forced to rely on the magnetic field, they try to calibrate their magnetic sense using other cues, especially the polarization of daylight at sunset, just before they start out."

Birds seem to possess uncanny skills as astronomers. Most birds migrate at night, Dr. Gould said. They recognize patterns of stars and the tilt and rotation of the night sky as navigation aids, as Dr. Steven T. Emlen in the United States and a German investigator, Dr. E. G. F. Sauer of the University of Freiburg, proved some years ago, Dr. Gould said. In the Northern Hemisphere, he said, the Big Dipper, the lip of which points toward the North Star, is probably an important guidepost for migrants.

Birds find their way not only in long-distance migration but also in relatively short-distance homing. These two activities probably involve very different navigational skills, biologists believe. For migration, precision of direction is less important than taking advantage of the winds and flying as economically as possible to conserve strength over immense distances. For homing, however, a pigeon or other bird must find its way to an exact spot 100 or so miles from its starting point.

"The jet stream carries charged ions through the atmosphere, and this causes rapid variations in the magnetic topology of the earth," Dr. Gould said. "These variations can be large after magnetic storms on the sun, and they can cause birds a lot of trouble in finding their directions on long flights."

But birds are very good at knowing when they are home, he said, and they seem to fly in somewhat the way children navigate when playing the game "You're getting warmer or colder."

Deprived of ordinary vision, homing birds nevertheless seem to sense when they are getting closer or farther away from their destinations. How they do this is not understood, but something apparently guides the birds.

"In some experiments, homing pigeons have been fitted with frosted lenses over their eyes so that they cannot see any details of the ground, and yet they still often manage to reach their nests from hundreds of miles away," he said. This certainly involves a consciousness of the exact magnetic topology and field strength of the home nesting area, he believes. Experiments in Germany and Italy suggest that smell may be an important homing cue as well, although many scientists in the United States discount this theory.

In the other *Nature* paper, German and Australian scientists headed by Dr. Wolfgang Wiltschko of the University of Frankfurt am Main presented strong evidence that the migratory Australian silvereye birds with which they worked can probably see magnetic fields. This startling conclusion is based on a series of experiments demonstrating that the birds can easily orient themselves according to the earth's magnetic field when their cages are illuminated by white, green or blue light. But in red light, the birds lose their ability to sense magnetic fields, and orient themselves randomly.

In a comment published in *Nature* about this paper, Dr. Gould said the experiment provided "dramatic evidence" supporting a theory advanced in 1977 by M. J. M. Leask of Oxford University in England, which was widely discounted at the time. Dr. Leask proposed that in some birds, at least, visible light enters the eye, impinges on the retina, and excites electrons in a pigment called rhodopsin, which is known to be essential to vision. If some of these electrons happen to remain in an excited state, according to the Leask theory, the pigment becomes "paramagnetic," a state in which it could be affected by magnetic fields.

The experiments conducted by Dr. Wiltschko and his colleagues do not prove this theory, but they seem to show that there is a link between vision and the magnetic sense, at least in Australian silvereye birds.

For most creatures capable of orienting themselves with respect to magnetic fields, however, the mineral magnetite is still believed to be the major sensory agent. It is present in bacteria that navigate by magnetic fields, as well as in the many vertebrates known to feel magnetic fields.

Dr. Joseph L. Kirschvink of the California Institute of Technology has determined that the human brain, like the brains of many other animals, contains tiny crystals of magnetite, a naturally magnetic mineral. He believes that this supports the controversial theory that magnetic fields may cause physical or chemical changes in the brain, including, possibly, cancer or other diseases.

"There is no reproducible evidence that human beings can sense the earth's magnetic field or orient themselves by magnetic cues," he said. "No one has replicated experiments conducted during the late 1970's by Dr. R. Robin Baker of the University of Manchester, England, in which he claimed that human subjects could orient themselves magnetically."

"On the other hand," Dr. Kirschvink said, "that doesn't necessarily mean that magnetic fields cannot be felt at some subconscious level, or that they have no biological effects."

In an interview, Dr. Kirschvink took no stand on the issue of whether magnetic fields emanating from power lines and electric appliances could be harmful to human health, but said, "you can't rule such effects out a priori on a physical basis." The ion channels of the nervous system, which control many physiological functions, may be influenced by magnetic fields of the kind generated even by household hair driers, he said.

—MALCOLM W. BROWNE, September 1993

Aircraft to Play Mother Goose
in Unusual Rescue Experiment

ONE MORNING recently, Bill Lishman led a gaggle of young Canada geese on their maiden flight. The bat-shaped wings of an ultralight aircraft carried Mr. Lishman aloft. The geese lifted off under their own untested wing power and for 10 minutes the birds and the man they believe is their parent circled over the rolling Ontario countryside east of Toronto.

"At first they don't have much endurance," Mr. Lishman said. "When the birds start to pant, you know they're getting tired."

The training flights are designed to prepare the geese for an imaginative experiment called Operation Migration. The birds will assemble in a V-formation behind Mr. Lishman's 250-pound aircraft, christened *Goose Leader*, and follow him 36 miles across Lake Ontario on the first leg of a 350-mile journey to a wintering place in Virginia.

If all goes well, in a few years Mr. Lishman, a 55-year-old Canadian, could use the same technique to lead the vanguard of a new flock of endangered whooping cranes on their first migration south from the Saskatchewan prairie.

"He's on the right track," said Richard Urbanek, a crane biologist with the United States Fish and Wildlife Service.

If cranes can be taught to follow an ultralight aircraft, this approach could be the best hope for creating a new migratory flock of these rare and stately wading birds.

In April, 138 whooping cranes from the only self-sustaining wild population flew north from their winter home on the Texas coast to breeding grounds in northern Canada. Another 15 whooping cranes survive on the Kissimmee Prairie of central Florida, the site of a new experiment, which has been plagued by bobcat predation of the captive-bred birds, to start a nonmigratory population.

But an ambitious project begun in 1975 to start a whooping crane flock that would migrate from Idaho to New Mexico ended in failure. Biologists used a technique called cross-fostering, placing 289 whooping crane eggs over 14 seasons in nests of smaller sandhill cranes. The female whooping cranes, however, were sexually fixated on the foster species and refused to mate with males of their own kind. Only eight of the whooping cranes are left at their summer home at Grays Lake National Wildlife Refuge in southeastern Idaho.

Now, said Jim Lewis, the federal biologist who is head of the international whooping crane recovery team, the Canadian Wildlife Service wants to start a migratory population that would breed on wetlands along the Saskatchewan-Manitoba border that historically were used by whooping cranes. And Mr. Lishman's technique, Mr. Lewis said, "has definite potential" for that venture.

Six years ago, Mr. Lishman, a sculptor who has a 100-acre bird sanctuary and aerodrome near Oshawa, Ontario, realized a boyhood dream to fly with birds by raising a dozen goslings and teaching them to fly behind his home-built aircraft, a motorized biwing glider.

The process, called imprinting, was first studied in the early 1930's by the renowned Austrian ethologist Konrad Lorenz and his colleague Oskar Heinroth. Geese hatched from eggs in isolation, they found, would follow the first large moving object they saw—in this case, their keepers—as they would have followed their parents. For the rest of their lives, imprinted birds tend to regard humans as members of their own species.

Unlike most songbirds and shorebirds, which are genetically programmed to migrate from their breeding grounds to wintering grounds and back, waterfowl and cranes must be led. Ducks, geese and swans must be led on their first flight south by experienced adults. "Young birds traveling with old birds pick up the cues—the sun, stars, landscape and earth's magnetic field—and so the migration map is perpetuated from one generation to the next," said Frank Bellrose, a waterfowl biologist with the Illinois Natural History Survey.

Young marsh ducks such as mallards stay with their mother only until they are ready to fly; the adult male plays no role in rearing the ducklings. Young canvasbacks and other diving ducks are abandoned several weeks before they can fly. "Juvenile ducks eventually join large aggregations, which include seasoned migrants," Mr. Bellrose said.

But in the case of geese, swans and cranes, young birds migrate for the first time with both parents as a family unit. They spend the entire winter together and return to their original nesting place the following spring before the group breaks up as the adults begin a new brood and the year-old birds venture off on their own.

The idea of Operation Migration was that young geese and swans could learn a migration route by following a surrogate parent, Mr. Lishman in his ultralight aircraft. The project's own takeoff, however, was delayed by red tape. At first, wildlife officials in both countries questioned the scheme's scientific merit.

At daybreak last October 19, with all the necessary permits in hand, Mr. Lishman was airborne at the head of a wedge of 18 Canada geese, destination Airlie Center, a private conference center in Airlie, Virginia. He flew a sleek French-made ultralight aircraft with a 34-foot wingspan and a 28-horsepower rear-mounted engine. "That's more than adequate for flying at goose speed, about 30 miles per hour," he said.

Behind the flock was Joe Duff, a photographer in an identical aircraft dubbed *Goose Chaser*. A rescue boat followed the strange formation across Lake Ontario. "There's always a nagging fear that the engine could quit on you," Mr. Lishman said.

At the first stop, at a farm airstrip near Albion, New York, west of Rochester, the birds landed and were herded into nylon pens. Bad weather delayed the second hop for three days, and while private runways along the route to Airlie had been scouted in advance, stiff head winds forced an unplanned landing in a Pennsylvania field. The geese and their human companions reached Arlie on October 25.

Flying alone, said Dr. William Sladen, a retired professor of ecology at Johns Hopkins University who is the project's scientific advisor, the geese could have made the trip nonstop if they had known the way. An ultralight aircraft, however, carries fuel for only five hours of flight under perfect conditions. The geese wintered at Airlie Center's ponds under Dr. Sladen's care, but plans for Mr. Lishman to lead them back to Ontario last April became academic when 16 of the birds, wearing identifying neck bands, left of their own accord and a dozen of them arrived at his doorstep.

One unanswered question is whether the geese will return to Airlie on their own. The odds are good. "From banding studies," Mr. Bellrose said, "we know that waterfowl go back to the same areas on the flyway as they do on their breeding grounds."

This fall's trip with a large flock of geese will provide an opportunity to observe at close quarters whether harnesses on radio-tagged birds affect their behavior. And it will be practice for an attempt to duplicate the feat next year with young sandhill cranes raised from eggs provided by the Fish and Wildlife Service.

—LES LINE, August 1994

Tiny Transmitters Make It Possible to Track Peregrines Day by Day

FOR YEARS as they watched peregrine falcons fly south each fall, biologists could do little more than wonder what paths they took to reach their winter haunts in the tropics.

Now peregrines wearing tiny satellite transmitters on their backs, flying from as far north as Alaska to as far south as Argentina, are revealing their autumnal travels in detail. Researchers say these are the first day-by-day looks at any such winged migrations to the tropics.

As the picture of peregrine movements begins to emerge, researchers say the birds, true to their name, which means having a tendency to wander, are being found far and wide, in locales both expected and unexpected. Recently, in Providence, Rhode Island, the first of the two teams to use satellite transmitters on peregrines reported its findings at a joint meeting of several scientific societies, including the Ecological Society of America, the Society for Conservation Biology and the Association for Tropical Biology.

"This is a real breakthrough," said Lloyd Kiff, science director at the Peregrine Fund in Boise, Idaho, commenting on the work of the two teams. "I'm really dazzled that they're doing this."

Brian Millsap, a conservation biologist with the Florida Game and Freshwater Fish Commission in Tallahassee, said: "It's been a huge question mark. These results show how truly outstanding the technology is. It's really opening a door to a whole roomful of questions that we've had for years and that everybody is anxious to start answering. These data are priceless."

Researchers say the travels of the peregrine are of particular interest, as the bird, once in serious decline due to eggshell thinning caused by the pesticide DDT, is now being considered for removal from the endangered species list. Because pesticides potentially dangerous to the bird are still in use in other countries, researchers say they need to know where the birds go in their wide-ranging travels.

The majority of what was previously known about the movements of peregrines came from the time-honored method of banding, in which biologists put identifying rings on the birds' legs. With peregrines, researchers have to reach the tops of cliffs and then rappel down cliff faces to nests to band the chicks. If those marked birds are seen again, the location is noted and researchers add another point to their scattered collection of connect-the-dots data. Something like 1 in 100 birds will be resighted, usually only once, making for a sketchy picture at best.

Researchers have also used low-flying planes to track individual peregrines that have been outfitted with radio transmitters. Expensive and dangerous, the work is frustrating as pilots typically lose birds at the Mexican border, if not before. The problem is that the biologists must take the time to go through border checks, while peregrines, which can travel 500 miles a day, speed onward.

So in 1993, when satellite transmitters, previously restricted by their large size to tracking moose, elephants or storks, became small enough to slip over a female peregrine's slim shoulders, researchers jumped at the chance.

That summer, Skip Ambrose, an endangered-species biologist with the United States Fish and Wildlife Service; Mike Britten, a wildlife biologist with the National Park Service; and Dr. Patricia Kennedy, a raptor biologist at Colorado State University in Fort Collins, began putting transmitters on peregrines in their breeding grounds in Alaska. That same fall, Dr. William S. Seegar, a research scientist for the Department of the Army, and his team began putting transmitters on peregrines migrating through Assateague Island, off the Maryland coast, one of the birds' few known points of intensive passage. To date, the two teams have tracked more than 100 birds.

"In a few years, we already have more wintering locations and know more about their migration routes than in the previous 20 years of work," Mr. Ambrose said.

What researchers do first is lure these winged hunters to a tasty caged bird. When a peregrine tries to capture the meal, its feet become entangled in fishing line nooses. Researchers then attach the $2,500 transmitter, about the size of a D battery with an antenna protruding, using a backpack-style

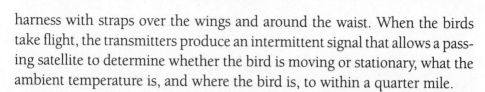

harness with straps over the wings and around the waist. When the birds take flight, the transmitters produce an intermittent signal that allows a passing satellite to determine whether the bird is moving or stationary, what the ambient temperature is, and where the bird is, to within a quarter mile.

As researchers trace these birds, filling in the large gaps left by the banding data and redrawing the boundaries of where peregrines fly, they are finding that the birds use numerous routes to numerous final destinations.

The team said recently it had tracked 17 peregrines from one study area on the Yukon River in Alaska to their wintering grounds in Brazil, Mexico, Honduras, Cuba and El Salvador, following one bird for nearly 9,000 miles. Some took unexpected routes through the Caribbean and along the eastern coast of Florida. One even stopped to spend the winter in Florida, something that came as a surprise. And even peregrines that are thought not to migrate are also being found.

Researchers track females, which weigh in at a little more than two pounds and can easily carry the three-quarter-ounce transmitter. It is a little too large for males, as they weigh closer to one and a half pounds.

Of particular interest, said Mr. Ambrose, is the finding that these birds, known to be fairly solitary and aggressive toward one another, do not appear to be congregating in any numbers when they travel either. The good news is that unlike other migrants, like shorebirds, which often gather in large numbers in a few places, this species is relatively safe from any isolated habitat destruction or concentration of toxins. The bad news is that it may be difficult or impossible to manage the birds, either in their travels or where they winter.

But there were a few sites, including an area on the western coast of Mexico near Culiacán, that researchers say several peregrines converged upon, enough to merit taking a second look to see if these could be gathering grounds that need managing. In fact, Dr. James Enderson, a peregrine specialist at Colorado College, in Colorado Springs, said researchers had already suspected the marshes in those areas, full of wintering ducks and songbirds, might be attractive to peregrines.

Earlier studies indicate that in North America, the bird reached a low of around 300 pairs in the mid-70's, with the complete extirpation of the species in eastern North America. By 1994, researchers reported a resurgence, to more than 1,800 known pairs.

However, the species has had varying success across the nation. As a result, Dr. Clayton M. White, an ornithologist at Brigham Young University in Provo, Utah, said the new research might prove especially valuable. Armed with the ability to follow individual birds, researchers could potentially track birds in areas with lower reproduction rates to see where they go and if their travels may be part of their problems.

Dr. Charles J. Henny, a research biologist with the National Biological Service, has already begun similar studies, along with colleagues, comparing peregrines' wintering migration routes and levels of pesticide residues in their blood.

—CAROL KAESUK YOON, August 1996

Songbird Migrations Tracked with Nocturnal Eavesdropping

BILL EVANS, a researcher at the Laboratory of Ornithology at Cornell University in Ithaca, New York, envisions a day when scientists and bird watchers will turn on computers on a spring or fall morning and, in addition to the latest weather maps, view charts showing the previous night's migration of songbirds bound for their breeding or wintering grounds.

In Mr. Evans's scheme, listening stations would monitor the nocturnal flight calls of thrushes, warblers, sparrows and other migrants. Computers would instantly identify every bird from its distinctive call and estimate how many individuals of each species passed over a particular site between dusk and dawn. One line of 40 stations, he proposes, would stretch 1,300 miles from Cape May, New Jersey, to eastern Nebraska.

Spring migration is the big event of the year for millions of bird watchers, who wait impatiently for northbound waves of songbirds that are often grounded by adverse weather conditions. "You could click the mouse on 'wood thrush' or 'Canada warbler' and see a map showing where flights of those species occurred last night," Mr. Evans said.

But his plan has a practical side as well. "Nocturnal flight-call monitoring could revolutionize the way we study birds," he said. "It has tremendous potential for tracking population trends as well as for mapping migration routes so conservationists can identify and protect critical stopover habitats."

Since 1966, the North American Breeding Bird Survey has used roadside counts of singing males to monitor population changes among 180 species of songbirds that nest in the 48 contiguous states and southern Canada. It was a 1989 review of survey data that first alerted scientists to population declines among migrants that breed in eastern woodlands and winter in Central and South America and the Caribbean. But many species nest in the roadless forests of northern Canada, and they are difficult to count using traditional methods. Scientists could detect population shifts among these boreal forest birds by studying tapes of their nocturnal flight calls over several years, Mr. Evans said.

His idea is more than a dream. Since 1991, the Cornell scientist has been eavesdropping on the fall migration across central New York state. From mid-August to early October, he maintains listening stations at up to seven locations along a 200-mile east-west transect, from Jefferson in Schoharie County to Cuba in Allegany County. Mr. Evans uses hearing-aid components to build pairs of weatherproof directional microphones that are mounted on the roofs of barns, houses and garages and are angled to cover a 75-degree expanse of sky. The sensitive microphones can pick up the calls of songbirds flying as high as 3,200 feet over a four-square-mile area. High-fidelity videocassette recorders are programmed to run for nine continuous hours, starting 30 minutes after sunset, to tape the sounds.

Mr. Evans has accumulated more than 3,500 hours of flight calls from the New York sites, and he has painstakingly analyzed many of those tapes, counting individual birds as their calls get louder and then fainter as they approach and pass the site.

For example, recordings from the night of August 28 to 29, 1993, reveal that large migrations of bobolinks, which nest in hayfields and old pastures, and a woodland thrush called the veery occurred in a broad front across the state. The largest bobolink flights, Mr. Evans determined, passed over Oneonta and Jefferson in the eastern part of the state. But the veery migration was heaviest at western stations in Alfred and Cuba. This suggests that factors like breeding densities may determine flight patterns, he said.

Most songbirds migrate after sunset, calling to each other with "chips" and "zeeps" and "sips" that on quiet nights are audible to the human ear. A skilled listener can identify the vocalizations of larger species, such as thrushes, cuckoos and the bobolink. Mr. Evans said the idea of recording their nocturnal flight calls occurred to him on a May night 10 years ago when he was camping on a bluff above the St. Croix River in Minnesota and counted more than 100 black-billed cuckoos, which are secretive birds by day, passing overhead in an hour.

"However, there are about 50 species of warblers and sparrows in eastern North America whose flight calls are so brief, less than one tenth of a second in duration, and so high in frequency that it is very difficult to identify them by ear," he said. Software developed by the Laboratory of Ornithology's bioacoustics research program has enabled Mr. Evans to identify and analyze these short calls on a field tape through computer-generated sonographs.

Cornell audio engineers are perfecting software that will automate recognition and counting of night migrants. "By the spring of 1997 we could have a pilot project monitoring such migratory songbirds as the American redstart and black-throated blue warbler as they arrive in Florida from their Caribbean wintering places," Mr. Evans said.

—LES LINE, October 1995

Bird-Watching Biologists
See Evolution on the Wing

THE WORK OF EVOLUTION is usually done on a time scale of millennia, too protracted for scientists to witness. So biologists were surprised by a finding that a bird known as the blackcap had evolved an entirely new migration pathway in a mere 40 years, right under the watchful binoculars of Europe's devoted bird enthusiasts.

The blackcap, a European wood warbler and relative of American gnatcatchers and kinglets, appears to have developed the genetic program for an entirely new migration route that shuttles it each winter from Central Europe to England rather than to the warmer climes normally sought by these birds in the western Mediterranean.

"The really surprising thing was that the adaptation to this new migratory behavior developed so rapidly," said Dr. A. J. Helbig, a post-doctoral researcher at the University of Heidelberg. He was an author of the study in the journal *Nature*. "It's the first demonstration that evolutionary processes can be much quicker than we thought," he said.

Researchers have been increasingly concerned about the ability of long-distance migrants to adapt to myriad changes in their environment as humans alter the planet ever faster.

"Long-distance migrants breed in one place, overwinter in another and pass through any number of habitats in between," said Dr. Frank Moore, an ornithologist at the University of Southern Mississippi. "Looking at this from the positive side, the paper shows the birds can respond rapidly. That's why this is so exciting."

But Dr. Moore cautions that while blackcaps seem able to adapt their migrating habits quickly, such evolution may not be a solution to all changes, particularly the habitat degradation commonly experienced by migrants.

"If the change is simply destructive," he said, "if a bird is given no alternatives, it doesn't matter how much genetic variability you have or how rapidly you adapt."

Researchers also expressed excitement over the study's support for the newly emerging idea that birds may be able to serve as accurate indicators of climate change.

Scientists say the blackcaps' migratory shift is typical of changes that have been seen in bird species around the world. For more than a century, ornithologists have noticed increasing numbers of species venturing farther and farther northward to pass the winter.

By showing that blackcaps were genetically programmed to fly to England, the study indicated that the recent worldwide changes in bird migration might be more than short-term behavioral adjustments to the weather. If the changes in blackcaps are any indication, Dr. Helbig said, the large-scale alterations in bird migration could likewise be genetic changes, the result of rapid evolution in action.

"The genetic evidence, that is the real news," said Dr. Paul Kerlinger, director of the Cape May Bird Observatory, a center for bird migration studies operated by the New Jersey Audubon Society. "We know that migration can change in very short periods of time, but being able to link those changes to a genetic component, these researchers have taken it two, even four or five steps beyond what's been done before."

Blackcaps are found across all of Europe, north to where the forest retreats from the cold, south to the Mediterranean and east to the Ural Mountains in Russia.

Before 1950, birds seen in Britain were either those there to breed for the summer before moving on to winter farther south, or those passing through during fall migration on their way from Scandinavia to sunnier climes.

Since that time, bird watchers have noted a steady increase in the blackcaps wintering in England. Today the birds can be counted by the thousands during the winter in England and ornithologists have found that many of the birds carry tags indicating that they spent the summer breeding in Germany and Austria.

To test whether a genetically altered migration program was causing these birds to deviate some 800 miles from the rest of the birds leaving Central Europe each fall, researchers from the Max Planck Institute in Radolfzell,

Germany, collected birds from Weston-super-Mare, about 100 miles east of London on the Bristol Channel.

In what Dr. Kerlinger praised as "ingenious, meticulous" experiments, Dr. Peter Berthold, Dr. Helbig and colleagues removed the birds to Germany and bred 41 experimental young birds over two seasons, along with birds caught in Germany for comparison. Since all the birds in the experiment had never been out of Germany and were subject to the same environmental cues, the researchers said, the only way for them to find their way to England would be through a genetically encoded map.

The birds' preferred migratory path was tested by putting them in covered cups lined with typewriter correction paper on which, as they tried repeatedly to take off, their feet scratched out their preferred direction in tiny birdprints on the painted paper.

The offspring of birds that had wintered in England left tracks in a direction northwest from Radolfzell, Germany, toward Weston-super-Mare, England. The offspring of birds that had wintered in Germany, by contrast, left tracks headed on the birds' standard route southwest toward the Mediterranean.

The researchers pointed to the birds' preferences for different migratory paths as evidence that they were guided by a map programmed into their genes. Researchers say that as much as 10 percent of the population that breeds in Central Europe and once went en masse to the Mediterranean in winter now heads instead for Britain.

Commenting on the study in *Nature,* Dr. William J. Sutherland of the University of East Anglia in Britain suggested that blackcaps might have been encouraged to pass the winter in Weston-super-Mare by the many Britons who put out birdseed. By fattening up the first few blackcaps whose genetic mutations brought them to England, bird lovers have given these creatures a better chance for breeding upon return to Central Europe in the summer.

The growing popularity of feeding winter birds both in America and Europe may be influencing migratory paths and populations worldwide, researchers believe. Those sunflower seeds and clumps of suet people put in their yards may thus in their own small way be influencing the course of evolution.

—CAROL KAESUK YOON, December 1992

7

PROTECTING AMERICAN BIRDS

Many bird populations are in serious decline as their habitats are changed by human development. In the United States the fragmentation of forests has exposed many songbirds to predation by the animals that are attracted by human settlements.

Valiant attempts are being made to restore threatened birds like the California condor and the thick-billed parrot by captive breeding programs. Despite the well-known success with the peregrine falcon, breeding programs are full of pitfalls and do not work in every case.

But the situation is not all bleak. The ducks of the prairie pot-hole region of the Dakotas and Montana are returning in record numbers after passage of legislation that protected wetlands.

Majestic Species' Fate May Ride on Wings of Six Freed Condors

ATOP A RED SANDSTONE escarpment half a mile above the rolling desert of northern Arizona, six California condors stretched their enormous wings and gazed out from a pen at the rugged scrubland, plateaus and piñon forests of their new home range.

If all goes according to plan, these majestic vultures will be released into the wild here 30 miles north of the Grand Canyon on Thursday as part of an intensive 15-year effort to save the species from extinction. It will be the first time scientists have tried to establish a colony of the condors, North America's largest bird, outside of California.

The event will mark a modest advance in the condor's recovery, and culminates nearly a year of sometimes bitter negotiations between biologists and local residents who feared the government would interfere with the area's ranching, logging and mining with overzealous measures to protect the birds from danger.

Fighting off a legal challenge by officials from San Juan County, Utah, in the state's southeastern corner, which delayed the project for several months, scientists have nearly completed acclimating the young condors to their new surroundings.

"I can't wait for one of the birds to catch a thermal and go almost out of sight," whispered Bill Heinrich, a biologist with the Peregrine Fund, a wildlife organization based in Boise, Idaho, that is managing the release.

"To see three or four of them circling overhead above the cliffs will be amazing," said Mr. Heinrich, who sat in a camouflaged blind near the birds' pen so they would grow accustomed to human contact.

A condor's wingspan can measure nearly 10 feet, making the bird a graceful glider that can soar up to 150 miles a day in search of carrion like deer, cattle and sheep. Adults have mostly black feathers, with orange heads, and can weigh 20 pounds. Juveniles are almost all black.

The condor's territory once included much of the United States and parts of Canada and Mexico. Bones dating back 10,000 years show that condors once nested near the release site, and anecdotal evidence put them in Arizona as late as the 1920's.

But human encroachment, mainly hunting and pesticides, caused the condor population to drop precipitously, and by the early 1980's only 23 birds remained in the world. To keep the condor from becoming extinct like the dodo, scientists captured the few remaining wild condors living in the coastal mountain ranges of Southern California and put them in an existing captive breeding program.

By 1991, their numbers had rebounded enough for scientists to begin a release program, which has been largely successful despite some conspicuous mishaps. Several condors were electrocuted when they flew into power lines (scientists now put fake power poles in their pens that deliver mild electric shocks on contact to teach them to stay clear) and one died after drinking antifreeze. There are now 17 condors in the wild, and 104 more in captivity.

Scientists believe that a second condor colony will help the species survive a catastrophe like disease or harsh weather that could wipe out a single population. Moreover, the condor's free-ranging ways inevitably put them near dangerous urban areas in California.

The ledges here on the Vermilion Cliffs, which jut from the plain in a sheer face for nearly 40 miles and are accessible by a rutted dirt road, are so remote that the birds will probably encounter only isolated ranch houses and tiny Indian villages.

"We think this is a better place than any we have tried," said Dr. Lloyd Kitt, the Peregrine Fund's science director and a former leader of the condor advisory board. "I'm optimistic. I will feel like a fool if they fly over that cliff and crash."

The two male and four female condors to be released are five months and six months old, and have never flown any distance longer than a flight pen. Young condors, rather than older ones that are set in their ways, were chosen because of the need to adapt if they are to survive. As a result of the condors' inexperience, Mr. Heinrich said, the release will probably be anticlimactic. The condors will probably fly to some higher roosts 50 yards from their pen, he said, pointing to a few large boulders, or just hop on the ground.

Not surprisingly, the condors will not be self-sufficient for some time, if ever. Scientists will lay out stillborn calves every few days for the condors to eat, gradually making the carcasses harder to find as a sort of hunting lesson before tapering off after two years. The feedings serve to keep the condors from wandering too far, and make them easier to track with radio transmitters fitted to their wings.

Condors prefer living on high ledges to catch the rising wind currents and to take advantage of nesting caves in the rocky crags. The released birds could eventually roam the canyonlands of southern Utah and the Grand Canyon.

Almost inevitably, they will alight on private land, which initially alarmed many Arizona and Utah residents who feared business-killing regulations. Residents of this region are known for their antigovernment sentiment, especially opposition to federal laws like the Endangered Species Act.

For example, many people blame federal logging restrictions for the closing of a lumber mill in nearby Fredonia, Arizona, in the early 1990's. The restrictions were imposed to protect the habitat of the Mexican spotted owl and the goshawk. Joy Jordan, a civic activist from Fredonia, was one of those skeptical about the reintroduction of the condors.

"Are we going to be put in jail if we hit a condor eating a road kill?" she asked. "And what if one drowned in a cattle trough? Would we have to pay a fine?"

To ease fears, officials from the Fish and Wildlife Service held several public meetings to explain a plan to designate the condor as an "experimental, nonessential" animal. A similar agreement was used to reintroduce wolves into Yellowstone National Park.

Under the plan, animals are considered expendable and do not affect land regulations or development. Wildlife officials also signed a letter with local governments that said they would remove the condors from the area if their nonessential status was changed.

The plan satisfied, or at least mollified, most residents like Ms. Jordan, who plans to attend the condors' release. But the Fish and Wildlife Service did not talk to officials in San Juan County, who later filed the lawsuit.

Bill Redd, a San Juan County commissioner who participated in the court challenge, said: "We know about the kangaroo rat in California, the riparian habitats on the East Coast and that nonsense about the spotted owl in Washington and Oregon. We are a rural county and we depend on the land."

It will be perhaps five or six years before scientists know whether the introduction of condors into Arizona has been successful. By then, the birds will be old enough to breed.

Watching the condors scratch and flutter in the holding pen, then lifting their heads as a golden eagle circled overhead, Mr. Heinrich said, "They're definitely ready to go."

—VERNE G. KOPYTOFF, December 1996

Forest Decline in North Cited in Songbird Loss

THE STEADY DISAPPEARANCE of migratory songbirds from North America's woodlands has been widely ascribed to deforestation of their winter habitat in Central America and the Caribbean. But a new study suggests that United States land development may be a much more potent factor in slashing songbird populations.

In a 23-year investigation of two species of warblers, Dr. Richard T. Holmes of Dartmouth College, Hanover, New Hampshire, and his colleague and former student, Dr. Thomas W. Sherry of Tulane University in New Orleans, found that deforestation of the birds' winter refuges in Jamaica did not account for the decline. Instead, they report, the warblers' worst losses, occurring in spring and summer, result from the steady encroachments of suburban development on the United States mainland. The findings are reported in a paper published as a chapter in a book called *Ecology and Conservation of Neotropical Migrant Landbirds* (Smithsonian Institution Press).

"In our censuses," Dr. Holmes said, "we found a strong correlation between the breeding success of a warbler population in one season and its overall population the following summer, regardless of how the birds fared in Jamaica during the intervening winter. Loss of winter habitat in Jamaica is a potential problem for the birds, but at present, it doesn't seem to be the major threat."

Conservationists generally accepted the results of the study. "Holmes and Sherry have done more work on this than anyone else, and I feel comfortable with their conclusions," said Deanna Dawson, a research biologist of the United States Fish and Wildlife Service at the Patuxent Wildlife Research Center in Maryland.

The scientists routinely counted all birds in their survey area, but focused on two species: the American redstart and the black-throated blue warbler. The two birds were studied and banded at both their summer breeding grounds in the Hubbard Brook Experimental Forest in the White Mountains of New Hampshire and in their winter area on the island of Jamaica.

The migrants lead difficult lives, living on average only two or three years, Dr. Holmes said. Despite their problems, the hardier birds, which weigh less than a half ounce each, sometimes survive far longer than their less-experienced flock mates, and the relatively rare older birds produce many more offspring. "There's one male warbler that we banded nine years ago, and he's turned up every year since," Dr. Holmes said. "That means he made eight round trips to Jamaica, a lot of flying for a little fellow."

The spread of humans into former woodlands has two dangerous effects on songbirds, naturalists agree. The first comes because predators accompany human populations. To determine the extent and type of predation on songbird nests, the scientists set up automatic cameras overlooking artificial warbler nests with eggs like those of songbirds.

"The eggs were raided constantly by a dozen or so different types of predators," Dr. Holmes said. "They included not only domestic and feral cats, but also raccoons, squirrels, chipmunks, skunks, blue jays and even black bears, most of them animals that live near human communities, feeding on garbage or handouts."

The second major threat to the species studied, and probably to most other songbirds, is parasitism by cowbirds, small members of the blackbird family, which, like cuckoos, lay eggs in the nests of other species. Cowbird fledglings, larger and more aggressive than their songbird nestmates, demand and get the lion's share of worms and caterpillars brought by harried foster parents. The number of surviving songbird chicks declines drastically. Cowbirds, like cats, congregate near human communities, especially farms. They are attracted by spilled grain and stay to use the free nursery service provided by songbirds.

"Studies in Illinois and other Midwestern states have shown that wood thrushes are basically raising nothing but cowbird chicks at this point," Ms. Dawson said. "So far, only about 10 percent of the wood thrush nests in the

Washington area have been parasitized by cowbirds. The problem varies regionally."

Loss of habitat is not limited to the felling of trees, Dr. Holmes said. The real problem is the fragmentation of large forests into housing lots that offer predators entry into the habitat.

Songbird populations do not suffer equally. "In the area we studied," Dr. Holmes said, "populations of some species increased somewhat and others remained stable, although over all, populations declined. In general, the birds did better during periods of major caterpillar infestation."

—MALCOLM W. BROWNE, November 1991

Snow Geese Survive All Too Well, Alarming Conservationists

THE STRIKING SIGHT of thousands of snow geese flying north in waving skeins across the spring sky is an annual thrill for mid-continent birders, and it should be, too, for environmentalists, who fret constantly about threats to waterfowl survival.

Instead, the ever-growing flocks of snow geese inspire deep conservation concerns, fears that after the voracious feeders reach their summer sub-arctic and arctic breeding grounds, thousands of square miles of delicate tundra will be destroyed, perhaps never to come back.

Thanks to an eclectic appetite and a brain that seems more flexible if not larger than might be expected, the snow goose has been so successful at adapting to human destruction of its southern winter home that the bird now threatens its summer home.

The goose population is now nearly three times as abundant as it was when first studied decades ago, and it is still growing. A colony that breeds at La Perouse Bay near Cape Churchill, Manitoba, and winters from the Gulf Coast to Kansas, had 2,000 nesting pairs in 1968. By 1990 it had grown to 22,500 pairs, an average annual increase of nearly 12 percent.

A team of prominent conservationists has proposed to halve the population by 2005, largely by reducing restraints on hunting throughout North America and by allowing the Inuits of Canada to gather their fill of eggs.

For the sake of the entire ecosystem—the delicate arctic vegetation and the other birds and vertebrates it supports—the conservationists say they have no choice but to reduce the number of geese that are fast tearing up the tundra. And the sooner the better.

Researchers estimate that more than three million geese, up from one million in the late '60's, spend the winter in the central states, and return each summer to the shores of Hudson Bay and Foxe Basin to breed.

But what seems to be a runaway success in conservation in the United States is fast turning into an environmental disaster where the birds nest in Canada. It is a disaster that could ultimately lead to the loss of other species, as well as the demise of countless geese and goslings.

"This is a very unusual problem, an overabundance of birds," said Dr. Bruce Batt, chief biologist for Ducks Unlimited, the largest waterfowl conservation organization in the world. "Our profession is not used to this kind of problem with migratory birds."

Dr. Batt is chairman of the international Arctic Goose Habitat Working Group, which studied the problem and has come up with some solutions that it believes fit with conservation goals.

Satellite photos taken in Canada over the years and analyzed by Andrew Jano, a specialist in remote sensing at the Ontario Ministry of Natural Resources, "helped us see the big picture, and it wasn't pretty," said Dr. Robert F. Rockwell, a professor of biology at the City University of New York and a research ornithologist at the American Museum of Natural History. "The photos show spreading rings of destruction by the geese. In the areas where the geese summer, 35 percent of the habitat is overgrazed, 30 percent is damaged, and 30 percent is destroyed."

Because of the short cold growing season in the arctic and subarctic, Dr. Rockwell said, it would take two or three decades for damaged habitat to come back. And because of sharp increases in the salinity of the soil, destroyed habitats may never come back, Dr. Batt said.

As a result of runaway consumption, known in conservation circles as a "trophic cascade," Dr. Rockwell said, "the geese are turning the tundra into a spreading slum." As they denude one area of edible vegetation, they simply move on to a place where the pickings are better.

"These are very opportunistic birds," said Dr. Robert L. Jefferies, a botanist at the University of Toronto who studies the interaction of the geese and vegetation. "We've marked young goslings before they could fly and found that they would walk up to 60 kilometers from their hatch site to exploit a new feeding area."

Dr. Rockwell, co-author of an article on the problem in *The Living Bird Quarterly,* a journal of the Cornell University Laboratory of Ornithology, traced the origins of this crisis from two phenomena of modern life that in more usual circumstances lead to the demise of species, not to population explosions.

Those are the loss of marshes along the coasts of Texas and Louisiana as a result of urban and agricultural development and the conversion of grasslands north of the coast into huge farms.

The marshes had been the birds' traditional wintering grounds, where a finite resource of reeds, roots and tubers in the brackish water controlled the size of the goose population. The limited nutrition in the vegetation could not sustain ever-expanding flocks. Not every adult could stash away enough nutrients to make the 2,000-mile trip north in spring and still have enough left to produce a brood of healthy, robust chicks.

But in the two decades after World War II, many of the coastal marshes were lost or severely degraded by urbanization. At the same time, farmers, aided in some cases by government subsidies, greatly expanded crop production on adjacent lands. The farmers in turn gave the birds a food subsidy.

"Now the geese had hundreds of thousands of acres of cropland to live on," Dr. Rockwell said.

With their serrated beaks, they had little trouble ripping out what farmers left in the ground to keep the land from eroding. The stubble and spillage of rice, corn and soybeans proved far more nutritious fare than *Phragmites, Spartina, Scirpus* and other marsh reeds, and the birds flourished on their new high-energy, high-protein diet.

The results, Dr. Rockwell said, included "a higher reproductive rate, a much higher adult survival rate, and offspring that were larger and in much better shape to survive."

As if that were not enough, federal and state agencies, with nothing but the best intentions, established wildlife refuges along migration routes to provide wetland habitats for breeding and migrating waterfowl. Many refuges provide corn and other grains to keep birds on the refuge.

And in the 1970's, lobbying by conservationists led to the establishment of "no hunting" zones in areas adjacent to the refuges, as well as strict limits on the hunting of geese generally.

Dr. Rockwell and his co-authors, Dr. Jefferies and Dr. Kenneth F. Abraham, a biologist at the Ontario Natural Resources Ministry, wrote that those measures "contributed to a nearly 50 percent reduction" in deaths among adult snow geese.

And because the geese breed for an average of eight to 10 seasons, surviving adults produced many more young, which in turn were more likely to grow up and reproduce successfully. And so the population grew in classic Malthusian fashion.

Climate, too, has played a role. Walter Skinner, a climatologist with Environment Canada in Toronto, analyzed about 200 years of climatic data for all of Canada. A temporary warming trend in the late 1960's and early '70's in the Hudson Bay–Foxe Basin region, where many of the geese nest, produced an earlier spring melt, permitting earlier nesting and increased reproductive success, Dr. Rockwell said. More recently it has been unusually cold in the high arctic, prompting migrating geese headed for northernmost breeding sites to spend extra time with more southerly colonies, adding to the environmental pressure at the southern sites.

"In 1984," Dr. Rockwell said, "100,000 extra birds stayed about five days at La Perouse Bay, and in that short time destroyed about six kilometers of coastal marsh."

The effects of heavy foraging on fragile arctic soil have been well documented in the 28-year study of the La Perouse Bay colony. Adult geese arrive each spring before the vegetation begins growing, Dr. Rockwell and his co-authors wrote. The birds feed by grubbing beneath the surface for roots and rhizomes of their salt-marsh forage plants.

That destabilizes the thin arctic soil and results in erosion by melting snow, spring rains and wind. Ponds form, growing every year as the birds grub along their edges.

On the remaining land, with the vegetation gone, evaporation increases, and salts from underlying sediments surface. As soil salinity increases, reaching levels three or more times higher than sea water, the forage plants decline, leaving behind few edible plants for goose meals and no live willow bushes for the geese to nest in.

The desecration is having effects on other species. At Cape Churchill, the yellow rail has declined, and Jim Leafloor, a biologist at the Ontario Natural Resources Ministry, has documented a drastic decline in the breeding of small Canada geese in areas where the foraging of snow geese has been intense.

The geese themselves are beginning to suffer from their success. Dr. Batt of Ducks Unlimited said there were early warning signs of a gradual loss in the health of adult geese and their ability to reproduce in some colonies.

"Now the goose population is so high," Dr. Batt said, "it is on a collision course, with destruction of the ecosystem surpassing its ability to support these birds, leading to a protracted decline. There is still time to do something about it. But we have to get on it right away, because we're losing ground every year."

Dr. Batt acknowledged that the solutions being suggested "won't be palatable to a lot of people."

"But," he added, "we have to kill birds, for their own good in the long run as well as for the good of the other birds and animals that are being shut out."

Ducks Unlimited rejected all solutions that would result in "wasting these terrific birds," he said. Instead, the group's suggestions would increase the use of snow geese for human food. The recommendations include extending hunting limits, allowing hunters to harvest three times more birds than now permitted and removing restrictions on the harvesting of goose eggs by the native people in Canada.

"Some of us are considering writing a cookbook of goose recipes," Dr. Rockwell said, noting that wild geese are much less fatty, albeit gamier, than domestically raised geese. "You can do everything with a snow goose that you can do with chicken," he pointed out. "It's really delicious."

—JANE E. BRODY, February 1997

Birds Rescued in Spills Do Poorly, Study Finds

THE IMAGE OF BIRDS mired in a sticky black froth is familiar after an oil spill. So is the scene of rescue workers rinsing birds' stained feathers with soapy water, nursing the sick to health and then releasing the shaken survivors into the wild.

Until recently, these intensive and costly efforts were considered relatively successful, saving thousands of birds from certain death. But new research shows that birds returned to nature after an oil spill, despite careful rehabilitation, typically die in a matter of months.

In a study published in the journal *Marine Pollution Bulletin,* Dr. Daniel W. Anderson, a biologist at the University of California at Davis, found that only 12 percent to 15 percent of rehabilitated brown pelicans survived for two years. In contrast, 80 percent to 90 percent of pelicans that were not known to have been exposed to oil survived as long.

"Brown pelicans are fairly tolerant birds," Dr. Anderson said. "We assumed they would do okay, but that is not what we found."

Dr. Anderson tracked 112 pelicans that were rehabilitated and released after oil spills in 1990 and 1991 off the coast of California. Using radio beacons and color markings, he found that nearly half were dead after six months.

Previous studies showed that many oiled birds died from a host of illnesses, like anemia, endocrine dysfunction and internal lesions. Birds frequently ingest toxic petroleum products during a spill, either by digesting the oil or breathing the fumes.

"Rehabilitated pelicans do fairly well right after release," Dr. Anderson said. "They're well fed and rested, but after a month my hypothesis is that they start suffering from immunosuppression and they cannot fight off diseases that they would normally encounter."

A bird's rehabilitation typically lasts from several days to two weeks and includes baths with a solution of detergent and water. Birds are given fluids for dehydration, sometimes a charcoal solution to counteract digested oil, and are put in pens with a pool of water to see if they float without difficulty before release.

One revelation in Dr. Anderson's study was that rehabilitated pelicans lose interest in mating. Only one of the treated birds spent time at a breeding colony, and that was just for a few days.

To some scientists, such dismal news about bird recovery undermines the worthiness of avian rescue efforts and raises an ethically charged question: Could the money spent on rehabilitation be better used for spill prevention and habitat restoration instead? Such a plan would mean leaving listless birds to die on oil slicks and coated beaches, a notion that some people find abhorrent.

While Dr. Anderson said it was premature to draw any conclusion about how best to deal with oiled birds, Brian E. Sharp, an Oregon ornithologist who has studied the survival of rehabilitated sea birds, has concluded that rescuing them is ineffective. In tallying liability damages for oil spills, he said a rehabilitated bird should be counted as a dead one. "We're doing this cleaning, which is supposed to be a fix of some kind and allows the public and politicians to ease their conscience," Mr. Sharp said. "But the birds aren't benefiting."

Mr. Sharp's research, published in the journal *Ibis,* showed that the average life expectancy of various rehabilitated sea birds since 1989 was four days. The average survival rate of oiled and treated common murres, white-winged scoters and western grebes was one fifth to one hundredth that of nonoiled birds.

For the study, Mr. Sharp examined banding and recovery records for sea birds living off the Pacific Coast from 1969 to 1994. He compared the date when oiled and nonoiled birds were tagged and when they were returned dead or injured.

Mr. Sharp found, perhaps surprisingly, that despite claims of advancements in bird rehabilitation techniques over the years, birds' survival had not improved during that time. For example, rehabilitated common murres survived a median of eight days before 1989 and six days after. Only two of 78 common murres lived more than a year after being oiled. "I was a little surprised," Mr. Sharp said. "I had always given the benefit of the doubt to the rehabilitation people. I thought probably they had done some good, but the data showed some very low survival rates."

The sea birds he looked at for the study are generally considered to be among the most fragile. Common murres, western grebes and white-winged scoters spend all of their lives on the ocean, and are reluctant to leave even an oil slick for shore. As a result, it can be days before these birds are rescued, during which time they are exposed to more toxic substances and are more likely to suffer a drop in body temperature due to feather damage.

The results of the two recent studies have not discouraged Dr. David Jessup, senior wildlife veterinarian with the California State Department of Fish and Game. He said Dr. Anderson's study did not reflect a quick response to oil spills. Rescue workers in the 1990 spill on which much of Dr. Anderson's data are based took nearly two days to build a care center. Dr. Anderson agreed that if the birds had not ingested so much oil, they probably would not have suffered as much.

"We're not ready to throw our hands in the air and say saving birds is a waste of time," Dr. Jessup said.

In a sign of hope, he said the percentage of birds surviving capture and rehabilitation had significantly increased. While only 2 percent of the birds rescued from a spill in 1967 lived long enough to be released, 60 percent survived treatment after a 1990 spill.

—VERNE G. KOPYTOFF, November 1996

Kirtland's Warbler Finally Finds Lots of Burned Forest to Call Home

NINE YEARS AGO, the Kirtland's warbler, one of North America's rarest song-birds, appeared to be on the ropes. A survey of its nesting grounds in north-east Lower Michigan turned up only 167 singing males, which translates into a roughly similar number of breeding pairs.

"We didn't know what it would take to bring the species back," said Harold Mayfield, an ornithologist in Toledo, Ohio, who is an authority on the bird. "I had my doubts about its survival."

But the warbler has come back in spectacular fashion after a forest fire created thousands of acres of the bird's specialized habitat: Christmas tree–size stands of jack pines. Now there are so many Kirtland's warblers that adult birds looking for nesting places have leapfrogged Lake Michigan and are apparently breeding in the state's Upper Peninsula. A census this summer counted 678 singing males, the second-highest number ever, in Lower Michigan. The survey also turned up 14 male warblers in four northern counties in the Upper Peninsula. At least six of those birds had mates, and observers saw adult birds carrying food, a sure sign that nestlings were being fed.

The sightings raise the possibility of a significant expansion of the species' breeding range, which has historically been confined to a few counties in Lower Michigan where there are large expanses of jack pines.

"We've got state and national forest lands with the right kind of ground cover that can be managed to favor Kirtland's warblers," said Ray Perez, a biologist with the Michigan Department of Natural Resources in Newberry in the Upper Peninsula. He noted that warblers had been found both in jack pine plantations and on tracts burned a few years ago by wildfires.

Mr. Perez said scientists would try to net and band the young warblers before they migrated to their winter range in the Bahama Islands.

Mr. Mayfield cautioned, however, that "exploratory efforts by birds to colonize new nesting areas often fail." He added, "I wouldn't be surprised if we can't find Kirtland's warblers in the Upper Peninsula three years from now."

Kirtland's warbler is a handsome bluish-gray and yellow bird about six inches long. It has a ringing song that can be heard for some distance, which is a big help to the biologists and volunteers who count the birds. The species is named for Dr. Jared P. Kirtland, a 19th-century Ohio physician and naturalist. The first specimen was collected on his farm near Lake Erie in 1851. But the bird's breeding grounds in the sandy, fire-scarred plains near Michigan's Au Sable River, a world-famous trout stream, remained a secret until 1903, when the first nest was found in a tract of jack pines that had been burned a few years earlier.

The jack pine is a northern tree that reaches nearly to the Arctic, but Kirtland's warblers nest only in the southern portion of its range. Jack pine depends on fire to propagate; its cones open and spread their seeds only after they are exposed to intense heat. The warblers appear about six years after a fire in areas where new pine growth is especially dense and the trees are five to six and a half feet tall, with live branches that reach the ground, Mr. Mayfield wrote in 1992 in his monograph on the warbler for *The Birds of North America* (American Ornithologists' Union and the Academy of Natural Sciences).

The warbler nests are built on the ground and hidden by shin-high vegetation, but the birds will abandon a site after about 15 years, when the slow-growing trees are 10 to 15 feet high and their lower limbs begin to die. "The females have very high real-estate standards," Mr. Mayfield said. "If the trees are too tall for a Christmas tree, they're too big for the warblers."

The extent of the Kirtland's warbler habitat and its population probably peaked in the 1880's and 1890's, Mr. Mayfield said. In 1871, he noted, an unchecked wildfire, fed by clear-cutting from logging operations that denuded most of Lower Michigan, burned one million acres in the heart of the bird's breeding range, converting the original white pine forest to jack pine. But by 1951, when conservationists took the first census, decades of fire suppression had reduced the amount of suitable habitat.

Moreover, brood parasitism by brown-headed cowbirds, which lay their eggs in the nests of other birds, had reduced the production of warbler fledglings to a critically low level. In one study, 70 percent of the warbler nests had been parasitized by cowbirds, and only two fledglings survived out of 29 nests.

Workers from the United States Fish and Wildlife Service, however, have killed more than 100,000 cowbirds from warbler breeding areas since a trapping program began in 1972, and parasitism has dropped, affecting 3 percent of the nests. State and federal foresters, meanwhile, have used clear-cutting, seeding, replanting and burning to replicate the effects of wildfires and create Kirtland's warbler habitats.

But biologists say the biggest help for the warbler came from a fire in 1980 near Mack Lake in Huron National Forest; that created 15,000 acres of new jack pine habitat. In recent years, nearly half of the breeding warblers have been found in the Mack Lake burn. Those trees, however, have reached a size where the birds are beginning to abandon the area. That is why this year's census was down from the all-time high, reached in 1995, of 759 singing males in Lower Michigan.

Foresters are planting 4,300 acres of jack pines this year in a scramble to make up the difference. Mr. Mayfield compared the drop in the number of breeding birds to a "bobble in the stock market."

—LES LINE, August 1996

Who Will Come to Duck's Rescue?
Ancient Forebears Are Enlisted

IN WHAT RESEARCHERS say would be the first use of ancient DNA in a recovery plan for an endangered species, scientists are proposing that a duck known only from the tiny island of Laysan be introduced to other Hawaiian islands where fossil bones containing its DNA have been found.

Conservationists say the new work, published in the journal *Nature,* is at the forefront of a movement to use paleontological evidence to justify reintroducing endangered species with highly restricted ranges to areas they inhabited in the distant past.

"I can't think of why this shouldn't be done," said Lloyd Kiff, science director at the Peregrine Fund in Boise, Idaho. Mr. Kiff, whose program is attempting to reintroduce the California condor to Arizona, where it has not occurred with any frequency for 10,000 years, said conservationists see this as a new way of protecting species.

But not everyone is enamored of the idea of sprinkling endangered species across the landscape, as they are seen by some as a possible threat to unencumbered human use of the species' newly claimed habitat.

The duck, *Anas laysanensis,* has been known in historical times only from Laysan, a 900-acre island 800 miles northwest of Kauai. The ducks inhabit a brackish pond in which they feed on brine flies and brine shrimp—highly unusual habits for a relative of the familiar mallard.

Biologists had assumed that the species had evolved to be highly specialized for its peculiar, if unstable, life on the tiny island. As a result, even though the species was known to be at risk from chance disasters like a 1993 drought, which fewer than 150 ducks survived, biologists did not consider reintroducing it elsewhere.

Helen James and Dr. Storrs Olson, paleontologists at the Smithsonian Institution, had been studying fossil deposits in lava flows on the major Hawaiian islands. Ms. James said they had recently become suspicious that some of the bones in these deposits might be those of Laysan ducks. Bones of such dabbling ducks, however, are difficult to identify definitively by size and shape.

Dr. Alan Cooper and Dr. Judith Rhymer, working with Dr. Robert Fleischer at the Molecular Genetics Laboratory of the National Zoological Park in Washington, used the bones as a source of DNA instead. Less than 3,000 years old, the bones used are known as subfossils as they still contain organic material and are not yet fully mineralized into rock. When researchers compared the DNA sequences from the bones with those of modern Laysan ducks they found they were identical.

The new study has radically changed scientists' view of this species by showing that it was once at home on Maui, Hawaii, Molokai, Oahu and Kauai from sea level to elevations of 5,000 feet. Researchers say they now think the pond on the island of Laysan is simply their last outpost and refuge.

The Laysan duck, like many other species in Hawaii, was most likely extirpated during colonization by Polynesians arriving some 1,600 years ago, either directly by hunting or indirectly by introduced animals or diseases. Researchers say they are hopeful that today, free from hunting and given a bit of protection and assistance, the Laysan duck could thrive once again on some of its former islands.

As with any reintroduction, the major biological risk is that the Laysan duck could bring with it new diseases from interim habitats. Another potential problem for maintaining the species is the risk of hybridization with game farm mallards, aggressive birds that have already formed hybrids with the native Hawaiian duck, or koloa.

Dr. Scott Derrickson, a biologist at the National Zoo's Conservation and Research Center in Front Royal, Virginia, notes that raising a captive population for release should pose no problems, as he and others have found that the Laysan duck will happily eat commercial duck food.

—CAROL KAESUK YOON, June 1996

Predicting Bird Extinctions from Deforestation

EVER SINCE the *Mayflower* dropped anchor in 1620, American settlers have migrated west, clearing the land of trees to make way for farms, homes and factories. Since then, nearly all the forests of the Eastern United States have been chopped down at one time or another.

Why, then, of the approximately 160 kinds of birds that inhabited the area from the Atlantic coast to the central plains and from Maine to Florida, have only four vanished? That constitutes a rate of extinction far below what ecologists might expect from such a vast loss of habitat. Critics have used the strikingly low number to challenge conservationists' claims that widespread deforestation in some parts of the world will inevitably lead to a severe loss in biodiversity.

In an article published in *The Proceedings of the National Academy of Sciences,* Dr. Stuart L. Pimm and Dr. Robert A. Askins examine the history of deforestation in the East and subsequent extinctions of bird species. Dr. Pimm, an ecologist at the University of Tennessee in Knoxville, argues that if only endemic species—those which live in the Eastern United States forests and nowhere else—are counted, the calculations of expected species loss hold up.

One reason, he says, is that the Eastern forest was never cut all at once; while the woodlands of Ohio were being felled by farmers moving west, for example, those cut earlier in New England were regenerating. Thus, there were always enough refuges for most forest birds.

But other biologists say that even with the revised counting method, forecasting extinction rates is an imprecise art.

The theory used to estimate species loss predicts that as an area of habitat is reduced, species will disappear at a predictable rate. If a habitat shrinks by half, for example, the theory predicts a 15 percent loss of species; a reduction of 90 percent of a given habitat would eventually cause half the species to disappear.

One of the originators of the theory, Dr. Edward O. Wilson of Harvard University, has said that it was intended to be a starting point. And indeed, scientists have begun to explore ways to improve its accuracy. The attention being paid to endemic species, or those that inhabit a solitary habitat, is one example.

Dr. Pimm and Dr. Askins, of Connecticut College in New London, estimated that at most, 28 endemic bird species inhabited the Eastern forests, including the four that went extinct. In the late 19th century, the period of greatest deforestation, half of all deciduous and coniferous trees were cleared. When all 160 bird species are factored in, the species-area relationship predicts that more than 20 species would have disappeared. But when the analysis is redone in terms of endemic species, the expected extinction rate is close to four, Dr. Pimm said.

"The story of Eastern North America is a remarkable experimental validation of what ecologists have thought all along," said Dr. Pimm.

Dr. Nicholas J. Gotelli, an associate professor of biology at the University of Vermont in Burlington, agrees that endemic species play a big role in predicting species loss. But he said that the model for predicting species loss due to habitat reduction is a crude tool, noting that it is often difficult to decide whether a species is found in only one area.

Dr. Bill Boecklen, an associate professor of biology at New Mexico State University in Las Cruces, is skeptical about making such precise predictions using the species-area relationship.

"They are right, it's the endemics that are most at risk," Dr. Boecklen said. "But the model was never built to consider that in the first place."

Other biologists contend that at least two extinct bird species—the passenger pigeon and the Carolina parakeet—succumbed to hunting rather than habitat destruction, a claim that Dr. Pimm acknowledges but does not believe detracts from his main point.

"More plausibly," Dr. Pimm wrote, "hunting was so effective because the habitat fragmentation concentrated the birds."

The picture in the tropics is not the same as in Eastern North America, says Dr. Pimm. "The density of endemics is at least 40 times, sometimes 100 times higher in the tropics," he said. "And that's of course what makes the rain forest so vulnerable."

—SARAH JAY, September 1995

Effort to Reintroduce Thick-Billed Parrots in Arizona Is Dropped

A SEVEN-YEAR EFFORT to return thick-billed parrots to the pine forests of Arizona where they once thrived has failed because birds raised in captivity floundered in the wild, quickly becoming prey for hawks.

Some of the birds starved, others succumbed to disease, but most were eaten by predators, often within 48 hours of their release. Researchers have suspended the project, saying they are uncertain whether any of the 88 parrots released in the Chiricahua Mountains of southeastern Arizona from 1986 to 1993 survived.

Reintroducing a dying species to the wild, even in its native habitat, is never easy, conservationists say. In fact, most attempts fail, especially when they involve animals born in captivity. Yet wildlife researchers say that such efforts tend to be more popular than other conservation techniques, like those that rely on legislation and public education.

To be sure, the appeal of returning the brightly plumaged thick-billed parrot to Arizona is largely esthetic, said Dr. Noel Snyder, an ornithologist who directed the parrot project and led the attempt to revive the California condor in the late 1980's. The thick-billed parrot, which is rather tame and not a prolific talker, is one of only two parrot species native to the United States. The other, the Carolina parakeet, went extinct early in this century.

The thick-billed parrot is emerald green with scarlet shoulder patches and a red nose. It disappeared from the southwestern United States in the early 1900's, a victim of hunters, but large flocks of the parrot still nest in the western Sierra Madre of Mexico.

The attempt to resurrect the species in Arizona began with 29 wild adult parrots that the United States Fish and Wildlife Service had confiscated from smugglers. Later, another 36 wild parrots were added. But the flocks dwindled considerably after a drought in 1989. Illnesses, such as an incurable, fatal wasting disease, took a toll as well.

The project encountered its most formidable challenge when it supplemented the wild parrot population with birds that had been raised in captivity. In their native habitat, thick-billed parrots are loud and gregarious creatures. Among the speediest parrot species, they streak through the sky, wingtip to wingtip, holding a tight, V-shaped formation. A robust thick-billed parrot can easily outfly goshawks and red-tailed hawks, its main predators, and sometimes even a peregrine falcon.

On the ground, the parrots seek safety in numbers, nesting close together in Ponderosa pine trees or Douglas firs at altitudes of 7,000 feet or more. Often, one bird will act as a sentinel for the others, perched on a treetop and calling vociferously when danger approaches.

The 23 parrots raised in cages lacked those flocking instincts and, more importantly, proved unable to develop such survival skills in the wild. Liberated, they took solo journeys to other mountain regions, foraged for pine cones in forests that had no pine trees and showed a lack of interest in socializing with their fellow avians.

"They get out there and the whole thing seems to be such an overwhelmingly new situation that they sit there dazed," said Dr. Snyder. "They don't flock properly, so they are very vulnerable to predation."

To improve the birds' chances, the biologists created what amounted to training sessions in how to behave like a thick-billed parrot. Housed in 30-foot-long cages with wild parrots, the ones raised in captivity learned through a combination of imitation and trial and error how to feed themselves.

"That's a tricky thing," Dr. Snyder said. They must snip a pine cone off a branch and then maintain a firm grip while using their thick bills to extract the seed.

The training helped hone the parrots' foraging skills, but it did not improve their flocking strategy. In one typical release, a female thick-billed parrot quickly located a wild flock and took up a perch about 33 yards away. When the flock returned to its feeding area, the female flew to another canyon. She was killed by a raptor later that afternoon.

"We reached a point where we were putting out the best candidates of captive-bred birds and just watching them die," said Susan E. Koenig, a volunteer in the program who wrote her master's thesis on thick-billed parrot behavior for the University of Arizona in Tucson. "Ethically, we couldn't justify putting out more captive-bred birds just to feed the hawks."

Wildlife Preservation Trust International invested $276,000 in the reintroduction program, some of which was a contribution from the Arizona Game and Fish Department. From the beginning, the project faced dismal odds. Only about one of 10 reintroduction schemes that use animals raised in captivity leads to self-sufficient populations of 500 or more individuals, said Dr. Benjamin B. Beck, associate director for biological programs at the National Zoological Park in Washington, who has reviewed 145 such projects.

The most successful reintroduction programs involved large numbers of animals over several years, Dr. Beck said. They also relied on public education programs to foster local support and enthusiasm.

"They capture the imagination," said Dr. Mary C. Pearl, executive director of Wildlife Preservation Trust International, a nonprofit organization based in Philadelphia. But she said most reintroductions occur when the species is in its last gasp. "That's almost certainly a prescription for failure," she said.

While the thick-billed parrot is not in such dire straits, conservationists say that a renewed North American population could become a critical safety net for the endangered Mexican birds, whose habitat is under siege by loggers.

In the next several years, Dr. Snyder, working with conservation groups in Mexico, will study that population. If they discover a healthy and sizable group of parrots, they might again try to rekindle a group in Arizona, but strictly with birds caught in the wild.

—SARAH JAY, May 1995

Decoys in Maine Lure Sea Birds, Gone a Century

DAY AND NIGHT, from last April to August in Rockland, Maine, bizarre noises blared over the enormous granite boulders that are heaped like dominoes atop Matinicus Rock, the emergent peak of a drowned mountain 20 miles off the coast here. From weatherproof speakers on the ledges came guttural bellows, shrill cries and a madhouse cacophony that on calm days could be heard a mile or more offshore.

The sounds, repeated thousands of times by a solar-powered compact disk player, were the voices of penguin-like sea birds called common murres. Lured ashore by the recorded calls of parent murres and their young, followed by the recorded roar of a huge murre colony, as many as two dozen of the black and white birds mingled with life-size decoys of adult murres and chicks and inspected ceramic murre eggs. They were the pioneers of what Dr. Stephen Kress, a research biologist with the National Audubon Society, hopes will be the first enclave of nesting murres on a Maine island in more than 120 years.

"The murres are prospecting for a new breeding site, but it could be several years before their numbers on Matinicus Rock reach critical mass and one bird takes the big step and lays the first egg," he said.

Dr. Kress, 48, has plenty of patience. As the self-appointed Pied Piper of the Gulf of Maine, he has dedicated his career to restoring sea birds on uninhabited outer islands where they were extirpated by 19th-century hunters. Due to the efforts of Dr. Kress and his teams of interns and volunteers over the last 21 years, Atlantic puffins are breeding once again on two former strongholds alongside thriving new colonies of Arctic, common and roseate terns.

207

Work with puffins and terns continues. Observers who spend the summer at spartan field stations on islands as small as seven acres count the birds every morning, monitor activities of nesting pairs and conduct feeding and productivity studies. "What they have in common is a passion for birds," Dr. Kress said of his paid and unpaid assistants. "And they have to like camping."

But the return of the common murre, which last bred off the Maine coast in the 1860's, is Dr. Kress's new priority. He calls his technique "social attraction." Recordings and decoys that are detailed right down to the fuzzy gray feathers of murre chicks and the multicolored splotches on hand-painted artificial eggs are a magnet for prospecting murres who in turn become living decoys that attract still more birds to the island.

Dr. Kress is using the same tactics to recruit a new colony of razorbills on Seal Island, seven miles from Matinicus Rock, where the species presently nests. The razorbill, a cousin of the murre, has a large beak and wings that seem too small to carry its stocky body aloft. Unlike murres, which typically drop their eggs on bare ledges, razorbills nest in dark crevices where they repeatedly utter a growl that sounds like a revved-up chain saw.

"For most of this century," said Dr. Kress, "there was a hands-off attitude toward sea bird conservation. Refuges were established, but nothing was being done to bolster declining species or to bring back birds that had been gone for decades. We're practicing hands-on conservation, learning how sea bird colonies are formed and developing techniques that can be used around the world."

Murres, puffins and razorbills are members of the auk family, two dozen northern sea birds that are the ecological counterparts of Antarctic penguins. Although penguins long ago lost the ability to fly, both penguins and auks are superb underwater swimmers, using their stubby wings as powerful paddles. Puffins dive as deep as 200 feet to forage for schools of small fish, but murres, with their greater body mass, reach depths of 600 feet. Since puffins and murres often prey on the same kinds of fish in the same waters, their different diving abilities enable the two species to divide the food resource without serious competition.

Penguins and auks share other traits. They stand upright on webbed feet, are long-lived, raise only one or two chicks a year and crowd together by the tens and hundreds of thousands on inaccessible islands where their eggs and offspring are safe from such land predators as foxes. The isolation of their breeding colonies, however, was not enough to protect the auks from human predators. From Maine to Labrador, their numbers were decimated by egg gathering, hunting, habitat destruction and the introduction of domestic animals.

Immense numbers of common murres once nested on islands off Newfoundland and Labrador, the epicenter of North America's murre population. But the birds were a source of fresh food, bait and oil for local fishermen. "By the late 1880's only very small groups of murres survived," said Dr. David Nettleship of the Canadian Wildlife Service.

Hunters by that time had turned their attention to tern colonies, supplying the birds' feathers to milliners. One party killed 1,134 common and Arctic terns on a single trip in Penobscot Bay, selling the skins for 30 cents each. In 1885 more than 75 thriving tern colonies were spread along the Maine coast from Saco to Cutler; by the end of the century, only 23 sparsely populated colonies remained.

Common murre and Atlantic puffin numbers in Newfoundland began a dramatic recovery in the 1950's. On Funk Island alone, Dr. Nettleship said, the murre population is estimated at 400,000 pairs. Three small islands within sight of each other in Witless Bay near St. John's account for 200,000 pairs of puffins.

But today there is only one small murre colony, with perhaps 125 breeding pairs, in the 900 miles of ocean and islands between Newfoundland and Matinicus Rock. Around 1980 the birds reclaimed Yellow Murr Ledge, a tiny rock outcropping off Grand Manan Island in the Bay of Fundy. Most of the puffin's southern outposts were never repopulated and only two colonies survived in the Gulf of Maine.

Terns, meanwhile, were protected after 1900 by Maine law and patrols by Audubon wardens. But the birds' recovery was stymied by herring gulls, whose numbers exploded in response to the proliferation of coastal garbage dumps where the gulls flocked to feed. Herring gulls nest on the same islands as terns and prey heavily on tern eggs and chicks.

Enter Dr. Kress and what began in 1973 as the National Audubon Society's Puffin Project. "There were only two puffin colonies in the Gulf of Maine and only one, Machias Seal Island at the mouth of the Bay of Fundy, had a substantial number of birds, around 900 breeding pairs," said Dr. Kress. "A single oil spill could wipe them out."

With the cooperation of the Canadian Wildlife Service, 954 two-week-old puffin chicks were transplanted from Great Island in Newfoundland to Eastern Egg Rock off the Maine coast in Muscongus Bay over the next 13 summers. The nestlings were placed in sod burrows, fed vitamin-fortified fish every day for a month by Dr. Kress and his workers and banded for identification before they left their nests. Decoys were planted on the boulders to greet them if they returned to the island after two or three years at sea.

The first human-raised puffins returned to Eastern Egg Rock, six miles east of Pemaquid Point, in 1981 and a few more birds arrived every year. In 1984 puffins, which breed at the age of five, nested on the island for the first time in a century.

Seal Island, now a national wildlife refuge, was the second historic puffin colony to receive transplants from Canada. Seven pairs were discovered nesting in 1992. This summer 19 pairs of puffins bred at Seal Island and 15 pairs fledged young at Eastern Egg Rock. "Seal Island is hilly, with deep soil and 100-foot cliffs, and it has the potential to be a huge puffin settlement," said Dr. Kress.

While waiting for puffins, the Audubon team eliminated nesting gulls from Eastern Egg Rock, scattered tern decoys in suitable habitat and broadcast tern courtship calls and colony sounds. "This was the first use anywhere of decoys and recordings for starting a sea bird colony," Dr. Kress noted. Thirty-two pairs of terns nested on the island in 1980, the first in 53 years. Despite setbacks, an avian cholera epidemic one summer and raids by predatory black-crowned night herons and mink, tern numbers have swollen to more than 1,300 pairs. Tern recolonization of Seal Island has also met with spectacular success.

Dr. Kress will not predict how soon common murres might breed on Matinicus Rock. Transplanting murre chicks from Canada, he explained, would not work. Unlike puffins, which are independent as fledglings, young murres are fed at sea by their parents.

"We've seen increased frequency of courtship displays, copulation and the passing of stones and fish between potential mates," he said. "Murres have even sat on artificial eggs while chick calls were being played. All the signs are favorable."

—LES LINE, September 1994

Prairie Ducks Return in Record Numbers

FEW BIRDS are more avidly sought after by nature fanciers and hunters alike than the ducks, and a lot of people have been distressed to watch the sharp decline of wild duck populations over the last decade.

Now, though, the biggest flocks in years are forming up on marshes across the mid–North American continent for their annual migration south. The once-dwindling numbers of mallards and widgeons, pintails and teal, redheads and canvasbacks, gadwalls, shovelers and scaups have suddenly and spectacularly rebounded, according to the United States Fish and Wildlife Service, raising hopes for a long-term recovery.

The population of breeding ducks in 1996 was double the average of the last four decades in the "prairie pothole" region of the Dakotas and Montana, the heart of the mid-continental "duck factory" where most of North America's ducklings are fledged—vivid evidence of how profoundly the restoration of lost habitat can affect wild creatures.

Under federal legislation passed nearly a decade ago, millions of acres of farmland in the prairie pothole region have been converted to grassland reserves where ducks can nest in safety from predators. At the same time, hundreds of thousands of acres of prairie potholes and other wetlands that attract and feed ducks have been restored. And two wet years in a row, after a decade-long dry spell, doubled the number of ponds and puddles in much of the pothole region during last spring's breeding season.

Together these three factors made the spring of 1994 the most successful breeding season for ducks since the 1950's and 1970's in the mid-continental United States and southern Canada.

The success confirms the operating premise of the North American Waterfowl Management Plan, a program in the United States, Canada and

Mexico to rebuild waterfowl populations over the long term, said a member of the program, Michael W. Tome. "When you put good habitat and water conditions together at the same time, you're going to produce ducks," said Dr. Tome, a Fish and Wildlife Service specialist in migratory birds.

But the permanence of the duck recovery is in doubt. The Conservation Reserve Program created by the Farm Security Act of 1985, the main purpose of which is to combat erosion on fallow land, rescued more than 10 million acres of farmland in the prairie pothole region. As it happens, the land is prime duck nesting habitat. But wildlife experts and conservationists are worried that Congress may not renew the program when it expires next year. They are also concerned that farmers may not sign up for a new program.

While "it's good to hear" about this year's resurgence of ducks, "it would be a mistake to say we've turned some permanent corner," said Dr. Joseph S. Larson, a wetlands biologist at the University of Massachusetts at Amherst.

Still, conservationists see this year's convergence of positive factors as a chance for further progress. It is "probably the best opportunity to produce waterfowl on a broad-scale basis we've had since the 1950's," said Jeff Nelson, chief biologist for Ducks Unlimited, a private conservation organization and leader in conserving duck habitat. "The reason is, we haven't had this extensive an area of grass with this kind of water since those days."

While wet and dry climatic cycles cause natural fluctuations in duck populations, few ducks will be hatched even in the wettest of years if there is insufficient nesting habitat. "Water attracts ducks to the landscape, but water doesn't make ducks," said Ron Reynolds, a wildlife biologist who heads the Fish and Wildlife Service's habitat and population study team based in Bismarck, North Dakota.

Different species of ducks make different use of ponds and wetlands. There are dabbling ducks and diving ducks. Dabbling ducks, also known as puddle ducks, feed in shallow water, dipping their heads just below the surface to eat plants, insects and small fish. They build their nests on uplands near the water. Mallards, pintails, teal, gadwalls, widgeons and shovelers are dabblers. Diving ducks like the canvasback, on the other hand, plunge into deep water to feed on the bottom and make their nests on floating mats of vegetation.

The dabbling ducks took a double hit when the wetlands were drained and their nesting areas converted to farmland. The loss, combined with a prolonged mid-continental dry spell, sent breeding populations into a tail-spin. The number of northern pintails shrank by nearly two thirds; that of blue-winged teal, widgeons and mallards dropped by nearly a third each, and the population of redheads by nearly a quarter. By the 1990's, North American ducks seemed in serious trouble. Most of the decline took place in the mid-continent, the very heart of duck country. The climate was more favorable and the habitat more stable on the eastern and western ends of the continent. But far fewer ducks breed there.

"The reason ducks have declined is low nesting success," said Mr. Reynolds. Diving ducks had fewer wetlands for their floating nests and the dabblers, confined to shrunken patches of grassland, could not conceal their eggs and young from foxes, crows and many other predators.

Governments have joined forces with private organizations like Ducks Unlimited and the Nature Conservancy to restore wetlands in the prairie pothole region, an especially important development for diving ducks. But the Conservation Reserve Program has brought the biggest habitat dividends by far. The program paid farmers to protect formerly cultivated land that lay fallow, often because of fallen crop prices.

Although the main purpose of the program was to prevent soil erosion, it turned out that as the grasslands grew back, an abundance of thick cover was created for ducks, as well as other birds like meadowlarks, black terns, bobolinks and sandpipers. Along with the grasslands and birds there returned a variety of wildflowers, small mammals and invertebrates.

The grasslands also tipped a balance between two of the ducks' main predators in the victims' favor. Foxes, which hunt a small area intensively and are the greater threat to ducks, gave way to coyotes that hunt a larger range and plunder fewer nests. "When we see coyotes on the landscape, we see that as being good news," said Mr. Reynolds.

At the same time, the biological enrichment that followed the regrowth of grasslands has been good for creatures like meadow voles and white-footed deer mice, offering predators a broader menu: "They don't have to key in on just duck eggs," he said.

The other good news for the ducks was heavy precipitation in 1993 and early 1994. Though the historic floods they caused wreaked havoc with the upper Mississippi basin, the summer rains of 1993 also filled many dry potholes and other wetlands. A snowy winter in 1993–94 further increased the number of spring ponds in prairie Canada and the north-central United States to six million, according to the Fish and Wildlife Service—47 percent more than a year before and a third more than the long-term average of 1955 through 1994. In Montana and the western Dakotas, the heart of the duck-producing region, the number of ponds was double the long-term average.

From this confluence of factors, more ducks congregated this year on the mid-continent breeding grounds than in many years. In a standard survey area encompassing the northern mid-continent and Alaska, the population of breeding ducks among the 10 most abundant species jumped to an estimated 32.5 million in 1994, up 24 percent from last year and 5 percent above the long-term average. The increase was mostly accounted for by Montana and the Dakotas, where the population roughly doubled between last year and this, and nearby southern Saskatchewan, where it grew by 64 percent.

Just as important, the hatching rate of ducks also went up this year. In the case of mallards, the most abundant and familiar species, about 30 percent of the eggs hatched, compared with 10 or 12 percent before the Conservation Reserve Program's benefits took hold, said Mr. Reynolds. "Mallards need a nesting success of about 15 percent to maintain their population," he said.

The Fish and Wildlife Service predicts that when the mallards that hatched last spring join their parents on this fall's migration to wintering grounds in the southern United States and the tropics, 12 million in all, 36 percent more than in 1993, will make the trip. Overall, the fall migration flight from the standard survey area is expected to total 71 million, 20 percent more than in 1993.

In fact, said Dr. Tome, this year's 32.5 million breeding ducks in the standard survey area approaches the North American Waterfowl Management Plan's long-term recovery target of 36 million for that area—roughly the average count of the 1970's, when at one point it exceeded 40 million for a record. For the entire continent, the recovery target is a breeding population of 62 million by the year 2000, with a fall migration flight of 100 million. No continentwide survey of the present population has been made.

The immediate concern, said Mr. Nelson of Ducks Unlimited, "is that we get them up to where they were in the 1970's before the next dry spell."

For now, anyway, pothole and pond resonate once more with predawn sounds that evoke images of abundance in the mind's eye. Aldo Leopold, the naturalist and conservationist, put it this way in his book *A Sand County Almanac:* "When you hear a mallard being audibly enthusiastic about his soup, you are free to picture a score guzzling among the duckweeds. When one widgeon squeals, you may postulate a squadron without fear of visual contradiction."

It will be almost like old times this October as squadron after squadron, species after species rises and, in Leopold's phrase, heads toward the equator "on quivering wings, ripping the firmament neatly into halves."

—WILLIAM K. STEVENS, October 1994

8

PROTECTING

BIRDS

ABROAD

Throughout the world forests are being cut down and grasslands converted to agriculture, changes that massively threaten both woodland and grassland habitats.

The shrike, for example, a key species of the grassland, is on the decline in many regions of the globe.

Even albatrosses, in their patrols of the distant oceans, are no longer protected from human activities. Thousands die every year in fishing nets, and DDT, still in use in many countries of the world, is building up ominously in the eggshells of the albatross populations of Midway in the mid-Pacific.

Exotic Birds, at Risk in Wild, May Be Banned as Imports to U.S.

THE STREAM of exotic birds imported to the United States will be sharply reduced if a bill passed by Congress is rigorously enforced. The bill covers almost all of the parrot species that are popular as pets.

Conservationists have sought for a decade to limit the imports on the ground that overharvesting of birds in the wild has decimated some species. Last year 400,000 birds, most of them born in the wild and many of them members of endangered species, were legally imported into this country, making the United States the largest importer of birds. The new bill would soon slash that number by at least half.

Imports of 10 bird species that are in immediate danger of extinction due to heavy international trading are to be banned immediately. Several hundred other species are to be included later. They are considered to be potentially threatened by trade by the Convention on International Trade in Endangered Species, an international treaty that has been signed by 130 countries, including the United States. The new law would also set guidelines under which some banned species could be imported in the future and would establish a program to help exporting countries conserve native species.

The legislation was negotiated by environmental groups, zoo associations, and representatives of the pet industry and bird breeders. It was approved by Congress late in the session with no opposition, and conservationists say they expect President Bush to sign it. A White House official said the Administration supported the bill in principle and it would be carefully reviewed.

"This is landmark legislation in wildlife conservation," said Ginette Hemley, director of wildlife trade monitoring at the World Wildlife Fund, an environmental organization in Washington.

The bill "is a rational way to regulate harvesting of wildlife species," said Marshall Meyers, executive vice president of the Pet Industry Joint Advisory Council, a trade association also based in Washington. "It cannot go on unfettered."

Six to 10 percent of American households now own a pet bird. Most of the birds sold in the United States, including the common canaries and finches, are bred in captivity and will not be affected by the legislation. Some 15 percent of the United States market, however, is made up of exotic imported birds. About 85 percent of those imports, including all but two parrot species, are caught in the wild, primarily in Indonesia, Tanzania, Senegal, Argentina and Guyana.

An estimated 6.5 million live birds were imported in the last decade. Twice that number may have been harvested from wild populations, since as many as 60 percent of captured birds are believed to die in transit.

Existing laws have proved powerless to stop the trade, said conservationists. Some species, like the Spix's macaw and the Bali myna, have already been reduced to only a handful of birds and are no longer traded because they are so rare. The Fischer's lovebird, one of the species covered by the immediate ban, is now rare throughout most of its range in Tanzania. Imports of these species will also be immediately banned: the yellow-headed amazon, the green-cheeked or Mexican red-headed amazon, the golden-capped conure, the gray-cheeked parakeet, the white or umbrella cockatoo, the Goffin's cockatoo, the lesser sulphur-crested cockatoo, the chattering lory and the red-vented cockatoo.

The problem is that the exporting nations are not properly carrying out the treaty, said Dr. Donald F. Bruning, chairman and curator of the department of ornithology at the New York Zoological Society. To trade in any species listed under the treaty, an exporting country must issue a permit declaring that the harvest of the birds does not harm the wild population. Many exporting nations do not have the resources necessary to issue scientifically based permits, said Dr. Bruning. Instead they issue quotas, which are then routinely ignored. Shipping birds with falsified documents is also common.

Except in the case of those species considered the most threatened, importing countries are not required to make a similar no-harm finding or issue any permits.

Under the treaty, Dr. Bruning said, "if another country issued a permit, the U.S. had to accept it." To halt the trade of a species, he said, "you have to prove it was decimating the species, and by then, it was heading for extinction."

Dr. Susan Lieberman, a specialist in international wildlife trade at the United States Fish and Wildlife Service, said, "We've been accepting everything on face value hoping that it was not detrimental to the species, and knowing it probably was."

Under the new legislation, the Fish and Wildlife Service will prohibit imports of most species listed under the treaty, regardless of any document issued by an exporting nation.

"What will be allowed is a very reduced trade under a lot of scrutiny," said Ms. Hemley.

Imports of captive-bred exotic species will be allowed, but foreign breeders will have to be certified by the Department of the Interior. It is expected that the legislation will spur new interest in captive breeding as the less expensive supplies of wild-caught birds dry up.

Exporters or other parties will also be allowed to petition the Secretary of the Interior to remove a species from the prohibited list. To gain an exemption, the petitioners would have to prove the wild-caught birds are being "sustainably" harvested, in other words that the removal of those birds will not deplete the population.

"In principle, sustainable harvesting is a good thing," said Dr. Steven R. Beissinger, a professor of wildlife ecology at Yale University. "But it is not easy with some species. It takes a few years of study, especially with a long-lived species like a parrot."

In one of the few studies of sustainable bird harvesting, Dr. Beissinger was able to increase the birth rate of a population of green-rumped parrotlets by making nesting boxes available. Theoretically, those additional birds could be harvested for export. But, he said, there is no proof that sustainable harvesting of birds will work.

In fact, said Dr. Beissinger, "there is no proof that any bird now in the trade is being sustainably harvested." By allowing species not yet on the treaty lists to continue to be traded, he said, "we are assuming sustainability when we have no evidence for it."

Dr. Lieberman said the legislation "won't solve the whole problem, but it's a step in the right direction," adding, "At least we are going to be able to stop and look first, and see if a species is okay, and we're going to be able to stimulate countries to come up with good management plans."

The smuggling of birds into the United States, which is a particularly serious problem at the Mexican border, is not directly addressed by the legislation. Conservationists speculate that smuggling might increase at first, but will probably drop off once there are fewer legal birds to provide a cover.

—CATHERINE DOLD, October 1992

Popular Pets, Vulnerable Species

Birds to be protected by import ban.

Agapornis fischeri (Fischer's lovebird) Scientific name (Common name)
    LEGAL U.S. IMPORTS IN 1989: 5,182
    APPROXIMATE RETAIL PRICE: $29–$39
    MAIN EXPORTING COUNTRY: TANZANIA

Amazona oratrix (Yellow-headed amazon)
    LEGAL U.S. IMPORTS IN 1989: 6,152
    APPROXIMATE RETAIL PRICE: $1,000–$1,500
    MAIN EXPORTING COUNTRY: NICARAGUA

Amazona viridigenalis (Green-cheeked or Mexican red-crowned amazon)
    LEGAL U.S. IMPORTS IN 1989: 0
    APPROXIMATE RETAIL PRICE: $1,000–$1,500
    MAIN EXPORTING COUNTRY: NICARAGUA, MEXICO

Aratinga auricapilla (Gold-capped parakeet)
    LEGAL U.S. IMPORTS IN 1989: 0
    APPROXIMATE RETAIL PRICE: $79–$99
    MAIN EXPORTING COUNTRY: ARGENTINA

Brotogeris pyrrhopterus (Gray-cheeked parakeet)
    LEGAL U.S. IMPORTS IN 1989: 6,385
    APPROXIMATE RETAIL PRICE: $29–$49
    MAIN EXPORTING COUNTRY: PERU

Cacatua alba (White or umbrella cockatoo)
    LEGAL U.S. IMPORTS IN 1989: 4,975
    APPROXIMATE RETAIL PRICE: $800–$1,200
    MAIN EXPORTING COUNTRY: INDONESIA

Cacatua goffini (Goffin's or Tanimbar cockatoo)
   LEGAL U.S. IMPORTS IN 1989: 5,917
   APPROXIMATE RETAIL PRICE: $600–$1,000 OR MORE
   MAIN EXPORTING COUNTRY: INDONESIA

Cacatua sulphurea (Lesser sulfur-crested cockatoo)
   LEGAL U.S. IMPORTS IN 1989: 712
   APPROXIMATE RETAIL PRICE: $1,000–$1,500
   MAIN EXPORTING COUNTRY: INDONESIA

Lorius garrulus (Chattering lory)
   LEGAL U.S. IMPORTS IN 1989: 1,100
   APPROXIMATE RETAIL PRICE: $79
   MAIN EXPORTING COUNTRY: INDONESIA

Cacatua haematuropygia (Red-vented cockatoo)
   LEGAL U.S. IMPORTS IN 1989: 6
   APPROXIMATE RETAIL PRICE: N.A.
   MAIN EXPORTING COUNTRY: PHILIPPINES

(Sources: Convention on International Trade in Endangered Species; U.S. Fish and Wildlife Service; Pet Industry Joint Advisory Council; New York Zoological Society)

Relative Puts Rare European Duck at Edge of Extinction

THE SEXUAL PROWESS of the American ruddy duck is threatening an endangered European relative, the white-headed duck, scientists say.

Europe's rarest duck, the white-headed *Oxyura leucocephala,* used to live throughout the Mediterranean region, but hunting and habitat destruction have wiped out most of the population since the beginning of the century. In 1977, only 22 white-headed ducks remained, surviving on three lagoons in southern Spain.

Spanish conservationists took action, banning hunting of the ducks and setting up more than 40 protected reserves and a captive-breeding program. The duck population has reached nearly 800.

Enter an unlikely threat: the North American ruddy duck, *Oxyura jamaicensis.*

Forty years ago, a bird lover, Sir Peter Scott, took three pairs of ruddies to a wildfowl reserve in England, where they flourished. Some of the British-born birds spread their wings, and about 3,500 wild ruddies now live in England.

A few hundred escaped to the Netherlands and France. And recently some have followed a British tradition and flocked to sunny Spain. There the sexually aggressive males are wooing female white-headed ducks. At least 10 fertile hybrids have hatched.

Conservationists say this interbreeding threatens to undo Spain's efforts to re-establish the white-headed duck in Europe and could pose a danger to the only other population, about 18,000 birds in Kazakhstan.

Spanish officials have started shooting the ruddies and hybrids, and have asked Britain to help solve the problem it created.

The Department of the Environment in Britain is financing a three-year, $100,000 study on how to contain the ruddy duck. Options include shooting the birds, clipping their wings, or pricking or boiling the eggs. An international conference was held last month to coordinate efforts.

"Nature put an ocean between these two species, and human interference brought them together," said Chris Harbard, a spokesman for the Royal Society for the Protection of Birds. "It's our responsibility to redress the balance and protect the white-headed duck, which faces extinction. The ruddy duck, which appears in large numbers in its proper native environment, does not."

More than 600,000 ruddies can be found in North America. Both birds are stifftail ducks. The white-headed duck has brown body feathers, a bulging blue bill, white cheeks and a mostly white head with a black cap. The ruddy is more reddish brown, with white cheeks, a less bulbous blue bill and a black head.

Mr. Harbard said this was the first time interbreeding between a native and an introduced species had threatened the survival of a bird in Europe, though a similar conflict has occurred in New Zealand, where introduced mallards have interbred with the native black duck, almost to the point of making the purebred black duck extinct.

The best way to protect the white-headed duck would be to deport the alien ruddy from Europe, but that may prove difficult. Since ruddies in Britain pose no threat to native ducks, they are protected there, and moves against them are likely to upset some bird watchers.

Dr. Myrfyn Owen, director general of the Wildfowl and Wetlands Trust in Slimbridge, England, said he hoped an accommodation could be reached.

"The main problem is the continued expansion of the ruddy duck," Dr. Owen said. "If we can reduce the ruddy to a reasonably small number in Britain, they shouldn't be a serious threat to the white-headed ducks."

Dr. Borja Heredia, an ornithologist who formerly worked for the Spanish National Institute for the Conservation of Nature, said the hybrids were a serious concern.

"It is not a habitat problem in Spain anymore," said Dr. Heredia, who now works with the International Center for Bird Preservation in Cambridge. "It's a problem of genetic deterioration.

"This isn't just a Spanish problem or a British problem," he said. "The white-headed duck is a native European species, so it is a European responsibility. If expansion of the ruddy duck is stopped, the white-headed duck will have a chance to recolonize not just in Spain, but in France, Greece, all over. If not, the white-headed duck is likely to become extinct in Europe."

—TERESA L. WAITE, April 1993

Mystery Surrounds Global Decline of Flying Robin-Size Predators

THEY CALL IT the butcher bird.

Named for its gruesome habit of skewering prey in a trophy-like array on sharp spikes and thorns around its territories, this keen-eyed hunting bird known as the shrike appears to be disappearing the world round. From the English heath to the Russian steppes to North America's grasslands, researchers are finding these birds to be in a precipitous decline.

Biologists say that what is happening to the shrike is symptomatic of the decline of grassland birds and the rapid disappearance of their flat, open habitat, which humans find perfect for development and farming. But though grasslands are rapidly being converted, researchers say they suspect that there is more threatening shrikes than simple habitat loss, though they are not sure what it is. As researchers shuffle hypotheses that point the finger at everything from DDT to fire ants, the birds continue to disappear.

"Overall, the picture is pretty bleak," Bruce Peterjohn, coordinator of the breeding bird survey for the United States Fish and Wildlife Service, said of North America's loggerhead shrike. "They're declining in all of their range. From the New England states they've pretty much disappeared." According to Mr. Peterjohn, as of last year, this once common bird was listed as extinct in Maine and Pennsylvania, endangered in 11 states and threatened in two others.

North America's loggerhead shrikes, as well as other grassland birds like Henslow's sparrows, eastern meadowlarks, field sparrows and dickcissels, have been on a steady decline for all 25 years of the breeding bird survey. In 1992 survey volunteers conducted 2,500 25-mile-long counts at the peak of the breeding season, tallying all the birds they could see and hear.

But more worrisome is the likelihood that these birds have been on the decline much longer, perhaps since the beginning of the century. In addition to the loggerhead, the only other shrike found in the United States is the northern shrike, which breeds in the boreal forests of northern Canada and Alaska and whose status is little known. Mr. Peterjohn said grassland birds had shown a more persistent and drastic decline than any other group of birds.

Dr. Reuven Yosef, a shrike specialist working at the Archbold Biological Station in Lake Placid, Florida, said the situation was the same around the world for many of the 70 species in the shrike family. "In Great Britain, at the turn of the century, according to their censuses, the red-backed shrike was as common as the blackbird," Dr. Yosef said. "But in 1989 the red-backed shrike was officially declared extinct in Britain. It's the same in Switzerland with the great gray shrike, with the lesser gray shrike across Europe. In Japan the bull-headed shrike and brown shrike are in terrible trouble. The story goes on and on."

Because of the growing concern, Dr. Yosef and Tom Cade of the World Center for Birds of Prey in Boise, Idaho, organized the first International Shrike Symposium, which brought researchers from 20 countries to the Archbold Station in January 1993.

The meeting confirmed fears that shrikes were doing poorly around the globe. But the world's shrike experts were unable to reach a consensus on what is behind the decline.

"It makes all of us fairly worried," said Dr. Carola Haas, a shrike biologist at Virginia Polytechnic Institute and State University. "The fact that they seem to be declining in all of North America is disturbing, and the fact that they seem to be declining worldwide raises the question of whether something special is happening to shrikes. No one really has a good idea why this is happening, and people are really now just grasping at straws."

The elusive nature of the disappearance of North America's loggerheads, one of world's best studied shrikes, is typical. Not unexpectedly, the disappearance of grasslands continues to harm loggerheads. But more puzzling to biologists is the fact that even in areas where the shrikes' habitat remains intact, their numbers continue to drop. Equally puzzling is the fact that the birds that do breed seem to produce in healthy numbers.

"There's habitat loss, but it's at a rate much lower than the losses of shrikes," Dr. Haas said. "The problem is that you can just look around and say this looks like really good shrike habitat but there aren't any shrikes here and so people have wondered, where are they? It makes you think something else is going on."

Some researchers say they suspect that problems come when the loggerheads, which breed from Canada south into Mexico, are all squeezed into the southerly half of their range to overwinter. As birds migrate south, encountering resident populations of shrikes, the numbers of shrikes living in an area have been seen to double. Given the general loss of grassland habitats, researchers say, the pressure on overwintering grounds to feed and maintain such high densities may be too great.

These robin-size birds, which are most easily seen sitting atop fence posts or snags, hunt by surveying the ground for a skittering grasshopper or even a small mouse or frog. Once fixed on their target, these hawklike predators swoop down and, lacking the talons of their raptorial counterparts, kill their prey with a swift and powerful blow of their head.

Also known to take other birds, these shrikes carry out their deadly work entirely in flight. Researchers say that like the hawks they so resemble, these predators may be vulnerable to poisons that their prey or their prey's prey ingest.

"Being at the top of the food chain and having an exclusively carnivorous predatory diet," Dr. Yosef said, "they accumulate all the negative things that we put into nature, fertilizers, PCB's, even benign things that we aren't aware could negatively affect the habitat."

Dr. Cade, who is also founding chairman of the Peregrine Fund, said: "Their hunting does make them particularly vulnerable to pesticides. They could have been done in quite likely by the more toxic agricultural chemicals used in the past like dieldrin and DDT."

Other researchers suggest that pesticides may simply have killed off too much of the insect prey base the birds rely upon. Yet other scientists have suggested that rivals like the voracious fire ant may be competing too successfully with these birds for food.

But scientists remain frustrated, unable to prove or disprove any of these hypotheses, and researchers are left with the difficult task of trying to save a bird without knowing exactly what is causing it to decline.

Researchers agree that the best understood and most obvious cause of shrike disappearances is the worldwide demise of their grassland habitat.

Because shrikes hunt like raptors, watching for prey crawling about in the grass, they require fairly well-grazed grasslands—prairie or pasture—to keep their food visible.

In addition to feeling the effects of urban development, Dr. Haas said, shrikes have been the victims of changing agricultural practices, as many farms continue to convert pasture land, which is ideal for hunting, to tall row crops, in which it is difficult for the birds to make their living.

In addition, as farming and ranching become more mechanized, the small bushes and trees that dot fields are often removed so large machines can quickly move through. These trees are essential as hunting perches and nesting areas for shrikes.

But too many shrubs and trees are similarly bad for shrikes. The loggerhead shrike has been particularly hard hit in the Northeast, where much land that was once kept in fields has returned to forests. While the return of forests has benefited warblers and other forest-breeding birds, it has slowly pushed out the many grassland species that had spread their range eastward over the last few hundred years. For conservationists, the return of the forests has raised the problem of which birds they should be trying to protect.

"It's a question that's being debated," Mr. Peterjohn said. "Here's a bird that expanded into this area in the past as a result of human changes in the habitat, and now subsequently they've changed again from what is favorable to shrikes to what isn't. How much management should we undertake for a species that wasn't native to an area? I don't think anyone has a solid answer for that."

While questions remain about the retreat of the newer Northeastern populations, conservationists agree that measures must be taken to protect the bird farther west within what is presumed to be its original range.

Dr. Yosef has already found simple ways to make unsuitable habitat much more inviting to these birds.

By adding additional fence posts on which shrikes could perch to hunt and guard a small territory, Dr. Yosef found he could enable many more shrikes to survive in an area. And though it is unsightly, a tangle of barbed wire provides a cozy nesting site for the birds, as well as ample barbs for storing their prey.

While these techniques will be useful in the management of shrikes, researchers say they cannot be counted on to save them. And the lack of solid answers about shrike declines has left scientists wondering, as have disappearing amphibians, whether these disappearances might not be a sign of a global environmental problem more worrisome even than the changes in their habitat.

"I see the shrikes as an indicator species," Dr. Yosef said. "And what this whole family of birds is showing us is that in some important way their habitats are in a lot of trouble."

—CAROL KAESUK YOON, March 1993

Old Nemesis, DDT, Reaches Remote Midway Albatrosses

Mick Ellison

AS ITS NAME SUGGESTS, Midway atoll in the North Pacific is a long way from anywhere: 3,100 miles from Los Angeles, 2,400 miles from Tokyo and 1,150 miles from Honolulu at the other end of the Hawaiian Islands chain. But isolation has not protected oceanic birds nesting on Midway from environmental contaminants that originate in distant places. Researchers have found high levels of DDT compounds, PCB's and dioxin-like compounds in black-footed albatross adults, chicks and eggs.

They have also found the telltale effects of exposure to these chemicals, which remain in the environment a long time. Among those effects are deformed embryos, eggshell thinning and what the researchers say is a 3 percent drop in nest productivity. The results are cause for concern because of the potential for greater harm to albatrosses and other species. PCB's and dioxin-like compounds have caused embryo deaths and chick deformities like crossed bills and club feet in double-crested cormorants, Caspian terns and other fish-eating birds that nest in Great Lakes colonies. The pesticide DDT and its breakdown product DDE gained notoriety 30 years ago in North America when they were blamed for collapsed eggshells that led to population crashes in birds like bald eagles, brown pelicans and peregrine falcons.

One of the researchers, Dr. Paul Jones, an environmental chemist in New Zealand, published a paper in *Environmental Toxicology and Chemistry* on the Midway albatrosses. And Dr. James P. Ludwig, the head of the research team, presented an abstract of his findings at a meeting of the American Chemical Society in Honolulu.

"If these findings were seen in the Great Lakes they would not be unexpected," said Dr. John McLachlan, director of the Tulane-Xavier Center for Bioenvironmental Research. "Coming in the middle of the Pacific Ocean makes them all the more dramatic, and I think these observations are potentially very important."

"Concentrations of persistent chemicals in black-footed albatrosses on Midway are nearly as great as current levels in bald eagles from the highly contaminated Great Lakes," said Dr. Ludwig, a bird population ecologist. Laysan albatrosses, which also breed on Midway, have a significantly lower toxic burden in their blood and eggs, and scientists attribute this to the two species' different feeding habits.

Dr. Ludwig, who led an international research team to the atoll, said the pollutants in the birds' diet were probably coming from India, Southeast Asia and Japan. The discovery came as a surprise, since there was no previous evidence to suggest that high levels of toxic chemicals occurred in sea birds in remote tropical ocean areas. "We chose the Midway albatrosses as representative of a relatively pristine environment," said Dr. Rosalind Rolland, a conservation scientist at the World Wildlife Fund in Washington. The organization sponsored the project with financing from the United States Environmental Protection Agency, but the study was halted last year because of budget cuts at the agency, canceling a final season of field work and delaying publication.

A member of the Midway team, Dr. John Giesy, a toxicologist at Michigan State University, said: "While albatrosses feed near the top of the food chain, they forage on the open ocean far from continental pollution sources, and we expected to find very low contaminant levels. Our research demonstrates that global controls on the distribution of persistent, bioaccumulative toxic compounds need to be considered. The problem can't be approached on a country-by-country basis."

Black-footed albatross nest productivity on Midway has been reduced by about 3 percent because of the contaminants, Dr. Giesy said. So far this effect is nowhere nearly as serious as the accidental deaths of thousands of adult albatrosses in fishing nets. Nearly 4,500 black-footed albatrosses were killed in drift nets in 1990. "Albatrosses are long-lived birds, and the loss of a young female that could breed for another 30 years is more meaningful than an unhatched egg," Dr. Giesy said.

Dr. Giesy noted, however, that concentrations of PCB's and dioxin-like chemicals in black-footed albatross eggs were at a threshold where further deposits would be expected to cause adverse population-level effects. "This is especially true because a very small increment in the toxic dose can cause a steep response," Dr. Giesy explained.

Dr. Ludwig said that amounts of DDT compounds in black-footed albatross eggs were running just under two parts per million. "These are not trivial amounts when you consider that eggshell thinning in a number of birds has been shown to occur at levels of between two and three parts per million," he said. "We looked at 4,000 albatross eggs, and crushing at the pointed end appeared to be the cause of two thirds of the unhatched eggs. The effect of eggshell thinning is very slight from a population perspective, but it's a warning that we shouldn't put any more of this stuff out there. And there's no guarantee that the black-footed albatross is the most sensitive species."

Albatrosses are good birds to study, the scientists said, because they live for 40 years or longer, mate for life, return year after year to the same nest and have no fear of humans. Most of the 14 species nest on islands in the Southern Hemisphere and during the nonbreeding season, they wander the southern oceans for months without once coming to land.

But the black-footed and Laysan albatrosses breed mainly on outlying atolls and islets in the Hawaiian archipelago, and they roam North Pacific waters as far as Alaska, California, Taiwan and the Bering Sea on wings that span six to seven feet.

The worldwide population of black-footed albatrosses is estimated at 100,000 birds, the Laysan albatross population at about 2.5 million. The two small islands at Midway atoll, Eastern and Sand, have a total land area of only two square miles but they support 7,000 nesting pairs of black-footed albatrosses and 200,000 pairs of Laysan albatrosses.

North Pacific albatross colonies were decimated by Japanese feather hunters early in this century, when 300,000 birds were slaughtered in one year on Laysan Island alone and colonies on Wake, Johnston and other islands were wiped out. The Midway colony recovered from the depredations of hunters by the end of World War II, but 54,000 adult albatrosses were killed by United States Navy personnel from 1955 to 1964 to reduce bird-plane collisions at a military air base on Sand Island, and black-footed albatross numbers on the atoll never returned to their 1945 level. The air base is now largely inactive.

Each albatross pair raises a single chick, which is fed partly digested fish and squid along with fat-rich oil from the stomach of parent birds that often forage several days' flying time from the nest. Albatrosses typically feed by sitting on the water and seizing prey on the surface. Flying-fish eggs floating on the surface are the main part of the black-footed albatross diet, but the birds also snatch flying fish as they flit past.

Laysan albatrosses, however, feed heavily on squid, which rise to the ocean surface at night. Concentrations of toxic chemicals are magnified in animals that are higher on the food chain, and Dr. Ludwig said that flying fish were half a trophic level higher than squid.

"This accounts for the higher levels of contaminants in the black-footed albatross," he said.

The use of DDT has been banned since the mid-1970's in the United States, Canada and Western Europe. But Dr. Ludwig, who heads an ecological consulting group in Victoria, British Columbia, said it was still the chemical of choice for control of mosquitoes and crop pests in developing countries. "There is a large, fresh DDT plume coming off the coast of Southeast Asia and currents are carrying it into the north-central Pacific, where the albatrosses intercept it," he said. "This is not old stuff that's been around for 20 years. The half-life of DDT is about three years, and the ratio of the parent compound to the breakdown product DDE in blood samples of adult albatrosses is about 1 to 1."

One likely source of the dioxin-like compounds, Dr. Ludwig said, is the large amount of partly burned plastic ingested by albatrosses. The birds are notorious for swallowing floating plastic debris, and virtually all of the plastic items found in Midway albatrosses originated in trash dumped on the coast of Japan and other Pacific Rim countries, he said.

—LES LINE, March 1996

Unusual Rescue Team, U.S. and Mexico, Tries to Save Rare Parrot

A TOWERING limestone cliff in El Taray, Mexico, that is the unlikely home to a rare species of parrot is being preserved by the Mexican government, with help from the United States government and environmental groups on both sides of the border.

The unusual joint response by the two nations could help protect the maroon-fronted parrot, an endangered, pigeon-size bird found only in a remote, high-altitude region of northern Mexico where the weather sometimes gets cold enough for it to snow. Nearly all other parrots live in warm or even tropical climates.

The parrots nest in crevices of the imposing cliff, and at times gather at the site by the hundreds. But there is concern for their future. The rugged, wooded countryside that has the cliffs at its center has recently become a popular site for vacation homes, and the clearing of the land—particularly the trees on which the parrots depend for food—is threatening the birds' survival.

The environmental movement in Mexico is about where it was in the United States 25 years ago, reliant primarily on government intervention without much public participation. Environmentalists hope that besides safeguarding the maroon-fronted parrot, the project here will set a precedent for Mexican citizens to get more active in helping to preserve threatened species and their habitats.

The cross-border action to protect one of North America's only parrots is a significant step because it combines American money, Mexican research and local support.

In September 1995, a Mexican trust established by the Environmental Ministry signed an $80,000 purchase agreement with a family in the state of Coahuila for 750 acres of highlands that include the limestone cliff that scientists have taken to calling El Taray (pronounced el tah-RYE), after the town of fewer than 100 people.

The United States Fish and Wildlife Service provided $55,000, but because of rules prohibiting the agency from buying land in foreign countries, the money was used for research. That freed a similar sum provided by a Mexican government agency, the National Council on Biodiversity, to be used for the purchase. The San Diego Zoological Society provided $20,000. The remaining $5,000 for the purchase price came from private sources. The preserve will be officially dedicated in 1996.

In all, about 24 cliffs within 150 miles of here have been identified as nesting sites for the maroon-fronted parrot, though some house only a handful of nesting pairs. Based on the observations of local residents and visiting scientists, the pockmarked cliff face here, at an elevation of 9,000 feet, serves as a kind of high-rise colony for about one quarter of all the breeding pairs.

Maroon-fronted parrots, which are mostly green with a dark maroon forehead and yellow eye rings, lay their eggs in small crevices formed over thousands of years by water seeping through the cliff. After hatchlings become fledglings, they and their parents are joined by other members of the flock in a raucous spectacle of sound and motion on, over and around the cliff. In October, scientists counted 1,500 parrots squawking and swooping around the rugged rock face.

Compared with some other bird species, little is known about the maroon-fronted parrot. But researchers say they have determined that the entire species numbers no more than 3,500 birds and that their range is confined to the northern Mexican states of Coahuila and Nuevo León.

"A lot is known about parrots in cages," said Ernesto C. Enkerlin, a professor of biology and sustainable development at the Monterrey Institute of Technology and Advanced Studies. "But, unbelievably, parrots in the wild are almost unknown."

Professor Enkerlin is the project field leader and will oversee scientific research at the preserve. The Museum of Birds of Mexico, in Saltillo, the nearby capital of Coahuila state, will administer the preserve.

The people behind the project hope that the parrots will become environmental ambassadors, easily recognizable, charismatic symbols like pandas that will motivate Mexicans to take more direct action to protect the diverse natural resources that are under increasing attack by development and overuse.

"This could be a very winnable conservation battle, because this is a species that people will identify with and care about," said Noel Snyder, director of parrot programs at Wildlife Preservation Trust International, a conservation organization based in Philadelphia that is helping set up the preserve at El Taray. That public identity with the parrots will provide "a big advantage in helping preserve not just the parrots but the habitat they rely on," Mr. Snyder said.

The scientists have already identified a few unusual traits of the parrots that must be studied before decisions can be made about protecting their habitat. The piñon pine tree cone is such an important a part of the birds' basic diet that the parrots have adapted their life cycles to the growing season of the tree. Nearly all parrots breed in the spring, but the maroon-fronted parrot breeds in the summer, when piñon cones are most plentiful to feed a gathering—and growing—flock.

Professor Enkerlin said the maroons were closely related to another temperate-climate parrot, the thick-billed parrot, which lives in the western Sierra Madre of Mexico and ranged as far north as Arizona until hunters killed them off there early in this century.

The thickbills rely on piñon pines not just for food but for nesting sites, using woodpecker holes or natural cavities in mature trees. But a scarcity of rain in the maroons' habitat stunts the growth of the pine trees and few trees reach a size suitable for nesting. This forces the birds to find shelter where they can, in this case in the cavities on the cliff face here.

Like other parrots, the maroons mate for life and are believed to live as long as 40 years. It is not clear whether the same birds return to the same site each year to breed, but Professor Enkerlin assumes they do.

The very remoteness that kept the cliffs from being discovered for so long now poses a serious threat to the parrots. The picturesque area is within a two-hour drive of the industrial cities of Monterrey and Saltillo, and the spectacular mountains and valleys are being carved up into vacation home lots.

On a recent visit, Professor Enkerlin was taken for a real estate prospect by a local family who wanted to sell him 300 acres of hillside as a development site. Later, when asked about the preserve, one member of the family,

José Inez Saucedo, said that he had heard that some people came in to feed the guacamaya, the Spanish word for macaw, which is what the parrots are called here.

"Who knows?" he said when asked whether people should make such efforts to help the parrots. "That's something they do, not us."

Until now, Mexico's National Council on Biodiversity never sought to preserve important habitats by venturing into land purchases or agreements with local agencies like Monterrey Tech or the bird museum in Saltillo. But the government has indicated that if the parrot preserve is successful it will set up other environmental trusts to help preserve species and habitats in other parts of Mexico.

For Professor Enkerlin, a former apple grower who remembers seeing the parrots without knowing anything about them, the project is a sign of Mexico's growing environmental maturity.

"A project like this probably would not have been possible 10 years ago," he said. "Now there are many more people involved in the environment to make it happen. We're still behind, but we're catching up fast."

—ANTHONY DePALMA, December 1995

Treaty Partners Study Fate of Birds at Polluted Mexican Lake

THEY CROSS the border in countless numbers, unprotected and at the mercy of their new host country. A few prosper. Most survive. Some die.

That is the normal cycle of life for the 250 species of North American birds that migrate from the United States and Canada, where they and their habitats are protected, to Mexico, where they generally are not. But that cycle was brutally interrupted last December when some 40,000 birds died on the fetid water of an agricultural reservoir near the central Mexican city of Silva, causing what wildlife experts say was one of the worst bird kills in North American history.

The incident at the Silva Reservoir, and the mystery surrounding its cause, were recently taken up as a test case by the Commission for Environmental Cooperation, created when the United States, Canada and Mexico signed an environmental side accord to the North American Free Trade Agreement.

Besides attempting to find out what caused so many birds to die, the commission will examine to what extent migratory species in North America can be protected by international law. By the time the commission issues its report, which is expected in September 1995, it will also have tried to define who is responsible for shared animal resources—like black-necked stilts from the United States and lesser scaups of Canada—that leave one country and die in another.

So far Mexico welcomed the independent commission. Mexico has had a poor environmental record in the past, and its leaders are just beginning to acknowledge that they have a responsibility to help protect wildlife.

If successful, however, the environmental commission has the chance to create a model for international cooperation that can go far toward protecting migratory wildlife and the habitats on which those species depend.

"Audubon's long-range goal is not just to bring this case before the NAFTA commission but to sit down with groups from all three countries to figure out how to prevent this from happening again," said Kathleen Rogers, wildlife counsel for the National Audubon Society. The society, along with two Mexican environmental organizations, the Group of 100 and the Mexican Center for Environmental Law, jointly filed a petition with the NAFTA commission concerning the ecological disaster at the Silva Reservoir.

Ms. Rogers said the representatives of all three nations have indicated a willingness to develop a comprehensive management plan for nongame species of North American migratory and local songbirds, shorebirds and sea birds. The plan would complement an existing agreement among the three nations that governs the 60 million to 100 million ducks and other game birds that can legally be hunted.

The commission has assembled a team made up of scientists from all three nations to determine what killed the birds at the reservoir. The researchers are in a race against time: the next migratory season begins in late October or early November.

The scientists have limited clues to help them piece together what happened last winter. There are no preserved samples of the dead birds, the reservoir itself has been emptied and earlier investigations produced competing explanations of what went wrong.

What the scientists will have to work with are the basic facts of the case.

The Silva Reservoir was created at the turn of the century to help irrigate agricultural land in the central state of Guanajuato, about 150 miles north of Mexico City. As in most of Mexico, it rains here only from June through September. The rest of the year hardly a drop falls.

Compared with the industrial mess in Mexico City, the area around León, in the state of Guanajuato, is an airy paradise. But in this century León has become the shoemaking capital of Mexico, home to about 800 tanneries that discharge their wastes directly into the Turbio River and small streams that feed the Silva Reservoir. Sewage wastes from León and other cities also end up in the Silva.

Despite that degradation, the 300-acre reservoir has become a major stopover for migratory birds on the Pacific flyway from central and western sections of the United States and Canada. Bird watchers in the area have documented more than 50 kinds of birds at the reservoir, including the white-faced ibis, green- and blue-winged teals, baldpates, northern shovelers and least sandpipers.

All of the scientists who investigated the kill concluded that although the reservoir is badly polluted, it took something more than the normal toxins to trigger the catastrophe.

Birds that drank the yellowish water from the reservoir or ate plants and fish that grow in it became sick within two or three days. All evidence points to an extraordinary pollutant discharge, but of what and from where? At first the Mexican National Water Commission, which controls the reservoir, believed it was chromium from a nearby chemical plant.

Later the commission changed its finding to endosulfan, a strong pesticide. Officials claim that sometime at the beginning of December, a blue Chevrolet truck was seen near the edge of the dam and that four people were seen dumping a vat of liquid into the water that turned red on contact.

Environmental groups in Mexico and later in the United States had trouble accepting the conclusion for several reasons. Autopsies on the birds last winter showed no evidence of endosulfan poisoning. Endosulfan is more toxic to fish than to birds, but fish in the reservoir were unaffected. And endosulfan is practically insoluble in water, so if someone dumped the pesticide into the reservoir, the poison would not have spread.

The environmentalists have also rejected a study by the National Autonomous University of Mexico, which identified a red dye as the probable culprit.

The environmental groups think it far more likely that the kill was caused by a combination of events that turned normal discharges into the Silva watershed into a fatal cocktail. They suspect local factories, most probably a chemical plant that produces huge amounts, or the local tanneries, which have powerful political connections to hide behind. The family of the former governor of the state, Carlos Medina Plascencia, owns one of the largest tanneries in the state.

In all, about 40,000 dead birds from 20 migratory and local species were buried near the reservoir.

Dissatisfied with the Mexican government's investigation, Mexican environmentalists joined forces with the National Audubon Society to make the Silva Reservoir the first case brought before the NAFTA environmental commission.

"We knew we had to intervene in this or the killing would just continue," said Homero Aridjis, president of the Group of 100. "The birds will come back in November even if the water is deadly."

The environmental groups did not take the most confrontational route, in which they would have had to charge the government of Mexico with failing to uphold its own environmental laws. Instead they filed their petition under article 13 of the commission's rules, which empowers the commission to study an issue in any of the three countries, come up with its own conclusions and offer a remedy that can prevent another kill in the next migratory season.

At a public meeting in León last month, Julia Carabias, Mexico's environmental minister, welcomed the intervention.

"It's going to help," she said in an interview after the meeting. "It's demonstrating that the commission will have a good role in helping the three countries cooperate so they can really advance in their internal environmental programs as well as in the ones we all share."

As part of their response, the Mexicans also plan to relocate the tanneries to a new industrial park with better sewerage, and to upgrade municipal waste treatment plants. However, similar plans made in the past failed and the businessmen have said they cannot afford to pay for the projects, which would cost an estimated $280 million.

The head of the NAFTA commission, Victor Lichtinger, said that the cooperative attitude of the Mexicans had persuaded him that the commission could play a useful role in the Silva case. It may recommend a monitoring plan and specific waste treatment upgrades. Mr. Lichtinger said the commission may also be able to provide money for initial studies.

"We are not as a commission setting the environmental agenda for the three countries," Mr. Lichtinger said in an interview, "but we certainly are setting a regional agenda." That agenda will focus on the importance of shared resources and coordinated efforts, although the responsibility for act-ing will in the end remain with Mexico, with the support and cooperation of Canada and the United States.

"Either we do it all together," he said, "or we don't do it at all."

—ANTHONY DePALMA, August 1995

A Former Resident of Guam Pins Survival Hopes on Another Island

IN THE 1770's, when the English navigator James Cook launched the scientific exploration of the South Pacific, flightless rails could be found on a number of tropical islands from the Solomons in the west to remote Henderson Island in the east. Two centuries later, the majority of those bird species were known or believed to be extinct.

Most of the lost rails were the victims of introduced predators: the Norway rat, feral cats and dogs and the mongoose. One survivor was the Guam rail, and in 1968 its population on the 209-square-mile island of the same name was estimated at 80,000 birds. By 1986, however, the Guam rail was extinct in the wild, exterminated not by alien mammals but by a voracious, alien snake.

In a last-minute rescue operation, the Guam Division of Aquatic and Wildlife Resources captured 19 rails in 1984 for captive propagation. Now a long-awaited attempt to return the Guam rail to the wild is about to begin. If biologists succeed, it will be the first time that a viable breeding population of a rare bird has been re-established after the species disappeared or was totally removed from its native habitat.

"Starting in late February or March, we will release 30 to 50 rails every three months for as long as it takes," said Dr. Kelly Brock, the biologist who directs the project for the Guam wildlife agency.

There are 105 rails in a captive breeding facility on Guam and close to that number at 14 stateside zoos. "Right now we keep the males and females apart because we're almost full," said Dr. Brock. Guam rails are remarkably prolific birds. A typical family group includes adults with a clutch of three or four eggs, one-month-old chicks and immature birds that begin breeding at the age of three or four months.

The Guam rail, though, will not be coming back to Guam, at least not in the immediate future. The species' new home will be nearby Rota, an island that apparently has not been invaded by its nemesis, the brown tree snake.

A strikingly marked bird measuring 11 inches from bill to tail, the Guam rail is a bit larger than the familiar Virginia rail of North American marshes. The flightless rails on South Pacific islands, however, were forest or grassland birds. The Guam rail feeds on snails, slugs, geckos and grasshoppers rather than crabs, amphibians and other aquatic life. And with effort it can flap three or four feet off the ground to snatch a fluttering butterfly.

Unlike other flightless rails, the Guam rail "coexisted nicely" with rats, cats, dogs, monitor lizards and indigenous predators such as the egg-robbing Mariana crow and Micronesian starling, said Dr. Brock. But the rail and other native Guam birds were doomed when brown tree snakes from Manus in Papua New Guinea stowed away on a Navy ship and invaded the island soon after World War II.

A skinny, tree-climbing constrictor, the snake feeds on skinks and geckos when young and then switches to warm-blooded prey. It can easily swallow an adult rail, said Robert Beck, a Guam biologist. By the time perplexed scientists decided the snake rather than pesticides or some exotic disease was responsible for the birds' disappearance, there were a million or more brown tree snakes on the island and hardly any native birds. Three native songbirds were wiped out, and the rail and a Guam kingfisher survive only in captivity.

Rota, 30 miles to the north and one fourth the size of Guam, is sparsely populated, and much of its original forest is intact, although wildlife officials are concerned about impending resort development. Biologists chose Rota because of its proximity to Guam and its similar ecology, and small releases of captive-raised rails were made on the island in 1990 and 1991. All of those birds disappeared within a few months, but scientists were not surprised because they were testing introduction techniques and sites.

There are critics of the choice of Rota as a release site. "Everyone agrees it's just a matter of time before the brown tree snake gets to Rota," said Dr. David Steadman, a rail expert at the New York State Museum in Albany. "The logistics would be more difficult, but there are uninhabited islands north of

Saipan where the probability of establishing a sustainable population would be a lot higher. I don't see it happening on Rota."

On Guam, researchers are experimenting with snake-trapping techniques and snake barriers. One plan would create a 90-acre forest enclosure where a small population of rails could nest in safety. Brown tree snakes would be repelled or killed by electrical wires on the front of a stone wall, explained Earl Campbell, a herpetologist.

Unless the snake is somehow controlled or eliminated, Dr. Brock conceded, it is unlikely that the Guam rail will again run free on Guam.

—LES LINE, February 1995

Acid Rain Is Cutting Birds' Reproduction in a Dutch Forest

ACID RAIN is causing a severe drop in the reproductive rates of songbirds in the Netherlands, scientists have found. In an environmental threat reminiscent of the widespread havoc caused by DDT and other pesticides more than three decades ago, the pollution causes changes in the food chain, leading the birds to lay thin and defective eggs that rarely hatch.

The problem may also be affecting birds in other parts of the world, the scientists said.

The study, published in the journal *Nature,* is the first to document causal links between acid rain and reproductive problems in forest birds, said one of the authors, Dr. Arie van Noordwijk, the head of the Department of Population Biology at the Netherlands Institute of Ecology, part of the Dutch Academy of Sciences. The disruption seen in the forest ecosystem, he said, is rooted in the levels of calcium found in soil, snails and birds.

The researchers studied birds in the Buunderkamp Forest about 12 miles west of Arnhem, which like many European forests has received significant levels of acid rain. They found that the percentage of great tits laying defective eggs had risen to 40 percent in 1988, from 10 percent in 1983.

Knowing that the soil there is naturally low in calcium, Dr. van Noordwijk and his colleagues suspected that further depletion of the soil because of acid deposits might be the ultimate cause of the egg defects. They set out to determine the chain of events.

The scientists first wanted to learn whether a lack of calcium was the cause of the eggshell defects. Just before the nesting season, they supplied some birds with additional calcium, placing feeders of snail shells and chicken eggshells in nest boxes. The birds that ate the supplements immediately began producing normal eggs, confirming that calcium was the missing factor. An investigation of the stomach contents and of nest droppings of normal great tits established that snail shells were the birds' preferred source of calcium.

The researchers then wanted to see what was happening to snail populations, which obtain calcium from their food and by ingesting soil and rock. Resampling old study sites, they found that since 1970, snails had seriously declined in forests where poor soils had been further leached of their calcium by acid rain, but had remained stable in forests with calcium-rich soils.

Adding calcium to poor soils by dumping tons of lime quickly resulted in a restoration of the snails.

All the links in the chain can thus be seen: acid rain, with its increased level of hydrogen ions, is causing a decline in soil calcium levels by leaching the chemical out as the water passes through. Snail populations are dropping in concert with soil calcium levels. And the birds, which cannot find enough calcium-rich snail shells to fulfill their needs, are laying defective eggs.

"This is not an isolated phenomenon that occurs in only a few areas," said Dr. van Noordwijk. "It could be affecting a significant portion of the world's birds." At least four other species in the Netherlands, including spotted woodpeckers, blue tits, coal tits and European tits, are known to be affected, he said.

Whether it is also affecting birds in the United States is unknown, American scientists said, because no similar studies have been done.

Dr. Stanley Temple, an ornithologist and professor of wildlife ecology at the University of Wisconsin, called the study significant, but said local effects, if they exist, "would be limited to forests that, like the Buunderkamp, have naturally calcium-poor soils." Such conditions are probably limited to Northern pine forests, he said.

Dr. Frances James, an ornithologist and professor of biology at Florida State University, said the findings could have significant implications for her studies of warblers in Eastern mountain regions.

"There has been a greater than 25 percent decline in some warbler populations," she said. "We don't know what is causing it, but the work in Europe suggests a possible connection that should be investigated."

The findings raised issues that go beyond the effects on bird eggs, said Dr. Michael Oppenheimer, an atmospheric physicist at the Environmental

Defense Fund, the research and advocacy organization in New York.

"This study underscores the subtle, complex and latent changes that occur in ecosystems due to acid rain," he said. "Even if you don't see this phenomenon here, it doesn't mean there aren't other effects happening."

Dr. van Noordwijk said the consequences of calcium depletion are "not yet as widespread as pesticide problems were, but they may well be worse in that the remedies are much less clear and will take far longer."

Planned reductions in the pollutants that cause acid rain, sulfur dioxide and nitrogen oxides, are not strict enough to halt the acidification of soil in Europe or in the United States, he and Dr. Oppenheimer said. In any event, Dr. van Noordwijk added, calcium levels will not recover without an extensive program of liming or other type of restoration. The only other possible route of recovery is that the birds will learn to adapt, scouting out chicken eggshells at picnic sites and farms, as some now do.

The problem, Dr. van Noordwijk said, is "likely to get worse."

—CATHERINE DOLD, April 1994

India Watches in Vain for Migrating Siberian Crane Flock

EVERY DAY, Arvinder Singh Brar raises his binoculars to scan the sky above the wetlands in Bharatpur, where India's best-known bird sanctuary is, and swings them across the ponds, marshes and acacia tree groves where tens of thousands of birds feed, flutter, rest, fly and call to each other.

A few moments later, Mr. Brar, the deputy chief wildlife warden at the Keoladeo National Park at Bharatpur, lowers his binoculars in disappointment: the majestic Siberian crane, for decades the star of the park's winged visitors from abroad, has not been sighted this season for the first time in more than 30 years, and ornithologists fear for its future.

In recent years, the number of Siberian cranes has steadily dwindled. In 1964–65, more than 200 of the visiting cranes were officially recorded at the 11-square-mile park. By the mid-1980's, the number had fallen to about 40 before dipping to an alarming six in the winter of 1991, then five the following year, and now none.

The cranes, with their striking orange and black faces and thin orange-colored legs, used to fly vast distances from their breeding grounds in the flood plains of the Orb River in Siberia to India.

Specialists believe that the war in Afghanistan, first between the Afghan guerrillas and Soviet invaders and later between feuding Afghan factions, took a heavy toll of the elegant crane, which is listed as an endangered species.

Wildlife experts said that many birds were shot in Afghanistan and Pakistan. There have also been reports of Afghans and Pakistanis hurling rocks wrapped in twine to bring the birds down.

The cranes that come to India are part of the western flock of Siberian cranes, said Mr. Brar, who comes from the northern state of Punjab and who is working with a tiny team of scientists and researchers from the United States, Russia and India to study the bird and consider ways of saving it from extinction. The eastern flock travels to China in the winter and is said to number about 2,900.

But it is the western flock, which historically flew to India in late October or early November every year before returning home between January and March, that is of concern to environmentalists and wildlife specialists. Over the centuries, experts believe, this flock has wintered in three major Asian wetlands: Iran, Pakistan and India, which drew the biggest number.

Crane specialists and researchers working here say that only six of the big white birds were sighted in Iran this winter and that none have been seen in Pakistan for several years.

"The western flock was never very rich, maybe about 300 strong, 30 years ago," said Subodh Chandra Dey, a senior official in the Indian Ministry of the Environment.

As the Siberian cranes went into decline, wildlife officials in the United States, India and Russia, with assistance from Japan, embarked on a project in 1992 to save the western flock and reproduce it in captivity.

Six young birds raised in the United States and Russia were released into the park. Ornithologists from the International Crane Foundation in Baraboo, Wisconsin, have been playing foster mother to help the newest batch, four female chicks, to adapt to the park. The older birds are male and are a year old.

Wearing a white-hooded gown and using a hand puppet in the shape of the head and neck of an adult Siberian crane, Meenakshi Nagendra of the crane foundation has been helping feed the chicks from a bucket and leading them out into the marsh to exercise and forage.

Indian officials and the scientists, who include a Russian, are considering ways that the young human-reared birds can replenish the western flock. One possibility, they said, is to house them in a zoo over the blistering summer; another is to develop a breeding center at Bharatpur itself. A third is to return two birds to the historical breeding ground in Russia, wired with satellite transmitters to help track them.

Earlier efforts to track Siberian cranes met with limited success: a Japanese satellite monitored one bird's movements to Afghanistan and then lost the signal. Another signal died in Siberia.

The transmitters were wired on common cranes and on the resident Siberian cranes here, too, in the hope that the two young adults might eventually migrate to Siberia with the common cranes.

—Sanjoy Hazarika, March 1994

Rare Bird Illuminates Bitter Dilemma

WHEN EDMUND SMITH saw a nondescript black and white bird that he couldn't recognize flying over the hood of his car in central Somalia, he had little idea that he had caught a glimpse of the Bulo Burti boubou, an extremely rare species of shrike previously unknown to Western science.

Working with Mr. Smith, a biologist, researchers in Somalia quickly captured the bird, still the only known example of the new species. But when they did, they found themselves in an ethical quandary becoming more and more common among biologists: to kill or not to kill.

Scientists out discovering new species have a long history of dutifully shooting, poisoning, drowning, crushing or otherwise doing in their finds to preserve them for future reference and study. The team of biologists who discovered the boubou (pronounced BOO-boo) bucked over 200 years of tradition. Instead of preserving their only specimen as skin and skeleton, they kept it captive for a year, then returned it to the wild, hoping it would help propagate its presumably beleaguered species.

When the robin-size bird flew off, it left behind a handful of feathers, some photographs, a few blood samples and an intense dispute about whether a very rare animal is more valuable dead or alive.

When a species is discovered, biologists normally choose one individual, the "type," as the standard that determines whether any other individual belongs to that species or to another. Because it was the only one they had, biologists chose the boubou as its species type. To systematists, biologists who specialize in discovering, naming and understanding the evolutionary relationships of species, the idea of letting the boubou type fly off was almost criminal.

255

The description of the new species was first published in *Ibis*, an ornithological journal, and was written about in *Trends in Ecology and Evolution*, a news magazine for scientists. The lone shrike has since engendered strong feelings and harsh words both in letters to journals and in conversations among biologists. While preservation-minded biologists have praised the release of the bird, others, especially museum systematists, call it shortsighted and overly sentimental.

As species after species approaches the brink of extinction, this choice has become more common and more pressing. Many new species are so rare that if biologists collect and kill even a single animal, they fear that they could actually push the species into extinction.

But other researchers contend that if scientists do not keep and kill their find, there will be nothing left with which to study the species but the fragmentary information and material that can be gleaned from a quick look at the live animal. Besides, they say, any species so close to extinction is doomed anyway.

Dr. Nigel Collar, research fellow at the International Council for Bird Preservation in Cambridge, England, was the conservationist who strongly advised the biologists in Somalia to release the boubou.

"I have no concern at all, absolutely no concern at all it was the right thing to do," he said. "It was totally and absolutely the right thing to do. We cannot possibly, as conservationists, advocate the collection and killing of a species right at the edge of extinction. That's not what we're in business for. One obviously feels sorry for some systematists that they don't have as much information as they might want."

But systematists like Dr. Storrs L. Olson, curator of birds at the National Museum of Natural History at the Smithsonian, reject the notion that taking one bird from a species could be enough to tip the scales from survival to extinction. He called this view pseudoconservation. "It's sentimentality getting in the way of good science," he said. "It's not rational. It's not logical." Most systematists say the release of a bird that defines a new species is a serious mistake. "We have standards," Dr. Olson said. "There's a reason for the standards."

And for systematists, unlike conservationists, studying an extinct bird, even a fossil bird, can be just as instructive as studying a living species. Dr. Scott Lanyon, head of the division of birds at the Field Museum of Natural

History in Chicago, said that it was time for systematists to take a stand. "If we don't respond to this kind of action, then others will feel that it's all right," he said. "This is a step backwards. There's a misconception out there that the birds are thoroughly known. There are lots of field guides out there based on museum specimens. Now people are asking, 'Since we've got the field guides, why do we need the specimen?' There've even been suggestions that once you've figured out what you want to know from specimens, why not just trash them? There's no way you can know today what you'll need to know about an animal a hundred years from now."

Dr. Richard Banks, a bird systematist with the United States Fish and Wildlife Service, says the trend away from traditional preservation is growing. "There were two or three instances within the last several years of people publishing photographs of birds, describing new subspecies, with nothing to serve as a specimen," he said. "I think that it's bad business, bad science. It's not science at all to describe a species on the basis that they did and without anything to serve as a type specimen."

In the case of the boubou, much is unknown and may remain so. The scientists were not even able to determine whether their bird was male or female. Unfortunately, most of the information biologists want cannot be gleaned from photographs or blood samples. Biologists interested in comparing the bird's bones with fossil skeletons, measuring its gut length or studying the details of the patterning of its feathers—the list could go on and on—are out of luck.

Even apart from this dispute, this bird has from the start seemed unable to avoid drama.

After capturing the boubou, biologists videotaped, photographed, tape-recorded and took a blood sample from the bird. But the blood sample was lost through an airline baggage mix-up while en route from Somalia to the biologists in Europe who could have analyzed it. Civil war in Somalia forced the caretaker of the lone shrike back to Europe, taking the bird with him to Germany. When it was returned to Somalia more than a year later the shrike could not go to the place biologists suspected it called home; they left it instead in the Balcad Nature Reserve, safer from the war than the shrubs around the Bulo Burti hospital grounds, where it had been found.

Eventually one of the boubou's genes was analyzed with DNA from feathers that were preserved in alcohol. And more blood for future analysis was taken from the bird and preserved. When the DNA data were analyzed and compared with the segment of DNA from other shrikes, the results confirmed what biologists strongly suspected when they first saw the bird: the Somalian shrike was different enough from the other known shrikes to be considered a new species.

The biologists named the bird *Laniarius liberatus* "to emphasize that the bird is described on the basis of a freed individual," they said in the *Ibis* article.

Dr. Lanyon said of the biologists: "I realize that these people see themselves as heroes. But that bird is almost certainly dead now anyway. If that bird is from the area that they captured it in, why would you ever want to put it out somewhere else? The best chance of having it find a mate, especially if it's so rare, is putting it back where you found it. This is supposed to be about species conservation, not individual conservation."

To a population biologist, says Dr. Lanyon, the idea that a casual observer could see the last of a species or even one of the last of a species is unlikely at best. Most species of small birds and mammals are not limited in their numbers by their abilities to reproduce or to find a mate but by the amount of habitat available to them, he said. These animals typically produce too many offspring to be supported by the environment, and removal of one or a few individuals should not affect the fate of these species in the slightest, he said.

Dr. Jared Diamond, a research associate in the bird department at the American Museum of Natural History, said this logic was flawed. "If the thing is in trouble, it's in trouble," he said. "But it would be absurd to say that everything in trouble is doomed." Dr. Diamond said that some species had grown back to strength from very few individuals. "The famous cases that come immediately to mind," he said, "are the Chatham Island black robin, in the New Zealand region, that came back from seven individuals, of which two were females and five were males. There was a breeding program launched and it's now back up to 100. And the Mauritius kestrel, I believe, it came back from one or two pairs.

"No conservationist is willing to give up on a species."

The dispute over whether to kill or to let live extends to other animals as well. It recently flared over an endangered shrew after two specimens were killed and kept. The discoverers had named the shrew

Crocidura desperata, "to point out the desperate situation of the new species."

"The days of shooting everything in sight as a means of identification are long past," said Dr. Charles Walcott, executive director of the Cornell Laboratory of Ornithology. "And if you'd got the last ivory-billed woodpecker you'd feel rather awful about collecting it. I think you're caught between a rock and a hard place, and these situations are only going to get more common."

—CAROL KAESUK YOON, April 1992

Two-Nation Effort Aids
Condors in Americas

AS COLOMBIA'S national bird, the Andean condor appears on everything from flags to soft drink logos. Finding it in the wild is the hard part. While hundreds of condors dotted the country at the turn of the century, by 1986 there were only 22.

But in the last few years, bird experts from Colombia and the United States have teamed up to save the Andean condor and a northern cousin, the California condor. In January, two California condors were released in the Los Padres National Forest north of Los Angeles, using techniques perfected in work with the more abundant southern raptor.

The Andean condors circling lazily above the glacial lakes at Chingaza National Park in Colombia began life in incubators in two California zoos and were sent to the Colombian Andes to help rebuild their species.

The first release in Colombia was in 1989 at the national park here, 35 miles east of Bogotá. Since then, a total of 22 captive-born birds have been freed, nine here, eight in the south at Purace National Park, and five at the Chiles Indian Reservation on the border with Ecuador.

The release of the two California condors in the Los Padres National Forest brought the program full circle; they are adjusting well to their new environment.

"We wouldn't have a California release program without the Andean condor," said Dr. Michael Wallace, who as bird curator of the Los Angeles Zoo directs the California condor breeding program. "Everything we do with the California bird has been first tested on the Andean."

Incubator births were among the first tests. Dr. Wallace and his colleagues at the San Diego Zoo bred captive Andean condors for chicks that could grow up to test the wilds of the United States for the later release of the California bird.

Those Andean chicks became the first condors raised by a hand-operated raptor puppet, to avoid direct human contact. At seven months old, 13 female chicks were fitted with lightweight radio transmitters and released in the California mountains.

The Andean species provided an ideal model because it is nearly identical to its California cousin both behaviorally and physically. And, although it is under siege throughout the Andes, the estimated total population is in the thousands, compared with the existing 52 California condors.

Chile and Argentina boast healthy condor populations, each with as many as 1,500. Peru has 1,000 at most, and Ecuador fewer than 50. There are no data for Bolivia. Venezuela declared the condor extinct in 1960.

Exact figures are impossible to ascertain because the birds all look alike and cover up to 150 square miles. Also, many mountain areas, controlled by leftist rebels and drug traffickers, are off limits to condor-watching scientists.

What is known is that the Andean condor has had its bones carved into flutes and its feathers plucked for Indian ceremonial capes, and it has lost its foraging ground to pesticides and progress in every country it once called home.

In Colombia, where many main cities are high in condor territory, the bird is particularly endangered. Because of this and because of ties with Colombian biologists, the American team decided to send its male Andean condors, ineligible for release in California, to the Colombian Andes.

The males were joined in October by eight of the females who had earlier tested the California wild. Of the 22 birds released in the Colombian Andes, only three have died, from bacterial infections, a death rate that Alan Lieberman, bird curator at the San Diego Zoo and coordinator of the United States program coordinator, calls "exceptional."

But just as important, said Mr. Lieberman, is the fact that the program is still surviving.

"When you consider that 25 years ago a condor overhead only meant target practice, and that people are now concerned with protecting them, you realize how far we've come," he said.

In Colombia, the program is managed by the Department of Natural Resources and financed by the Foundation for Higher Education, a non-profit conservation group. The yearly budget is just over $20,000.

The Colombian program was dealt a setback in late February when Juan Manuel Páez, one of three field biologists, was killed by unknown assailants who forced his car off the road and shot him in front of his two young nephews. Family members told the local press that the murder was a case of "mistaken identity." The case is under investigation.

Mr. Páez's co-workers in Chingaza have assumed the feeding and monitoring of the birds, but no one has yet been appointed to permanently replace the young biologist.

In an interview just a week before he was killed, Mr. Páez said that a large part of his job was going into towns and Indian reservations to talk about the condor and why it was important to protect it.

"As scientists, it's crucial that we be completely accessible and that we make the program understandable to everyone," he said.

He pointed to one sign of progress: some Indians now use paper models of condors in their rituals instead of the real bird.

—TWIG MOWATT, March 1992

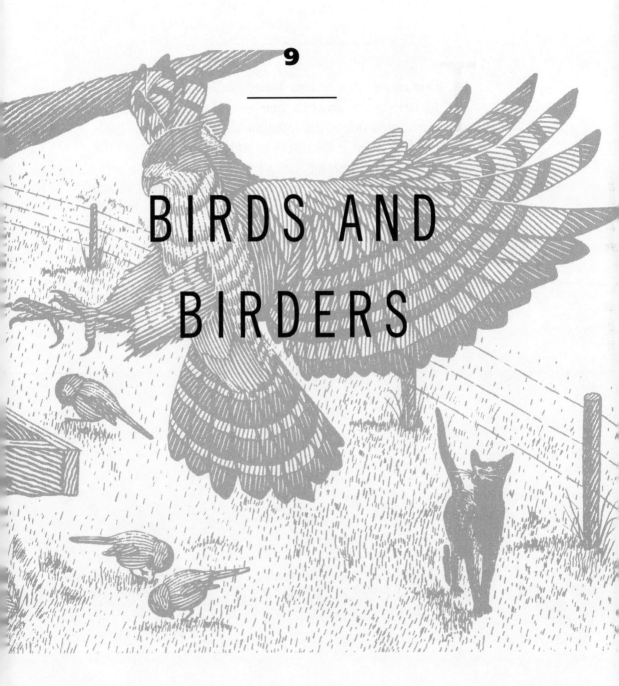

9

BIRDS AND
BIRDERS

There are people who have a strong relationship with birds. It's strictly one way but there are compensations for the unrequited side of the equation: those dedicated to birds know their labors enable the rest of us to enjoy or appreciate the refinements of avian behavior.

Ornithology can require its devotees to work until dawn, weather all manner of disappointment, and persevere through every kind of tide and terrain. The pieces that follow give a glimpse of birders at work.

Flocks of Biologists Take Wing to Michigan for Mating Season

BENEATH A BRILLIANT three-quarter moon in a grove of quaking aspen five miles south of Lake Superior, William L. Robinson, a professor of biology, was in search of the woodcock.

With a team of four volunteers, he set out to net and band the elusive game bird in its courtship grounds at the height of mating season.

He did not catch any; none that night in the wetland grove in Marquette County nor any the next cold night in a drier, older aspen grove, 40 miles to the southwest in Copper County State Forest.

In the early spring, the boreal forests of the Upper Peninsula of Michigan are alive with a symphony of spring peepers and the tramp of the biologists who come to study the mating and nesting habits of birds after a long, harsh winter.

The decline in populations of woodcock and some other woodland birds here and in other parts of the nation has been the subject of Dr. Robinson's work for the last 16 years. Other biologists in the area were studying game birds like the sharp-tailed grouse, and raptors like the goshawk and red-shouldered hawk.

Dr. Robinson was unfazed when the net result from setting five "mist nets" of fine nylon filaments ranging from nine to 19 feet in height was a single whippoorwill. "Maybe it's the moon," murmured the 63-year-old biologist, gazing skyward. Over the years he has netted and banded 517 woodcock and, in the midst of the wetland symphony, he had at least heard the low buzz of the woodcock's mating call, which is onomatopoetically called peenting. The team also heard the wingbeats of males spiraling high above the young forest in their annual courtship flights and spotted a few swooping past against the moonlight.

On an afternoon search, Professor Robinson was rewarded with the flushing of a female by a pointer and the discovery of three newborn chicks, otherwise virtually invisible with their brown-black streaked protective coloring amid fallen maple leaves and twigs. In a matter of minutes the three were weighed, averaging 28 grams, estimated to be two to three days old and banded.

Among the other biologists at work in the area was Stacy Christiansen, 25, a graduate student of Professor Robinson at Northern Michigan University. Trekking along the narrow roads in northern hardwood and pine stands of the Lake Superior State Forest, she sought to find three members of the Accipitridae family: the relatively rare goshawk and red-shouldered hawk, and the relatively common red-tailed hawk.

Armed with a cassette player, she played recordings of their calls every four miles to three points of the compass: the *keer-keer-keer* of the redshoulder, the sustained *keeeer* of the redtail and the strident *kak-kak-kak* of the goshawk. This is tedious work, three to five hours a day. In two and a half weeks she got responses from four redtails and two redshoulders, but no goshawks. She noted these along with careful observations of the habitats.

With binoculars she also sought out their nests, marking the trees and logging these along with sightings of other raptors—American kestrels, broadwing hawks, merlins and peregrine falcons—as well as the odd pine marten and coyote.

The work is part of a multiyear project on habitat protection for woodland raptors conceived by Professor Robinson and William W. Bowerman of Eagle Environmental Inc. It is focused on the goshawk, a candidate for federal listing as a "species of concern," and the red-shouldered hawk, which has threatened status in the state of Michigan.

The biologist out earliest, in the Hiawatha National Forest, an area of 892,000 acres, was Kevin R. Doran, 37, the wildlife specialist of the Forest Service based in Munising. Coming off three days of duty fighting wildfires, he was up at first light about 5 A.M. and on the lek, or courtship grounds of sharp-tailed grouse, 10 miles to the south, before six.

As the sun appeared on the horizon, the first of seven males appeared, softly churtling (the mating sound they make with their feet), glottally *coo-coo-coo*ing and prancing about, sometimes flying a few yards, always with their tails raised high. Females, invisible amid tufts of big blue stem and switch grass, observed the rite silently. "When their hormones are at a peak, they'll dance for 45 minutes," the biologist said of the males, pointing out one sharptail that had mounted a spiky white pine stump to show dominance as he performed his dance.

The sharptails seemed oblivious to the close-by sandhill cranes, vesper sparrows and some upland sandpipers just arrived from Argentina that are also of concern to conservationists. Mr. Doran, observing through binoculars from 50 yards away in the cab of his four-wheel-drive vehicle, said the sharptails had probably stationed a sentinel to watch out for predators like hawks.

His main mission was to count the male sharptails which, he said, "are species of special concern in Michigan" because of their low numbers. Unlike the more prevalent ruffed grouse, no hunting of the sharptails is allowed here in the Hiawatha.

Where he spotted the seven on this traditional lek, there had been twice as many males in previous years. He also logs the other species and keeps track of the grasses and other vegetation in this 1,500-acre savanna, which had been clear-cut by a timber company a decade earlier and subsequently turned into a field by controlled fires.

From this ground near Ready Lake, he drove north to a second-growth forest of beech and sugar maple a few miles below the shore of Lake Superior, where ice floes stretched northward for a mile. There a local firewood cutter had been dived upon a few days earlier by a goshawk, the biggest and most aggressive of Accipitridae.

Sure enough, as Mr. Doran neared a beech tree with a large nest in the fork of its biggest limbs, the mother goshawk let out a fearsome *kak-kak-kak* screech and swooped down within a few feet of his head—once, twice and a third time. "I could see her red eyes, and they were right on me," he said with a touch of awe. The trees around were immediately marked and there will be no more woodcutting in the vicinity, so as not to disturb the nesting female, he said. As wildlife biologist for the Hiawatha Forest, Mr. Doran is responsible for all its species, of which 225 are birds. But he is especially sensitive to the presence of goshawks because they are at the center of a dispute over the Forest Service's proposal to sell about 1,500 acres of second-growth trees in an eastern portion of the national land for selective and clear-cutting to a timber company.

A coalition of four environmental groups appealed that proposal in March under the stipulations of the National Forest Management Act of 1974, mainly on the contention that felling the trees would destroy important habitat for the goshawk, particularly older growth forest stands. Their submission persuaded the regional forest director in Milwaukee to remand the proposal the next month to Hiawatha's Munising office, where Mr. Doran is now busy rewriting the timber leasing proposal with fresh analytical data.

Environmentalists who have done a lot of research on the matter say there are a pair of goshawks and four nests in the disputed tract. It might also be habitat for a lynx or even a cougar, said Douglas R. Cornett, a conservationist who heads Northwoods Wilderness Recovery Inc., one of the groups that appealed the proposal to sell the land.

"In our opinion, the national forests are being grossly mismanaged," he said. "We don't have enough information to be harvesting national forests at the current rate. I would advocate a moratorium."

Mr. Doran said he resented the suggestion that he was less sensitive to the environment than those involved in the appeal. "But I am responsible for 325 species, mammals and birds," he said, "and I worry that by concentrating on a single species, the environmentalists may anger Congress so much that the legislators might ditch the entire Endangered Species Program. But that program has been a great success here. See the recovery of wolves and bald eagles, to name just two. And it bothers me to find people who think all timbering is evil when it is clearly beneficial to some species."

—DAVID BINDER, June 1996

Harsh Lessons of Survival
at the Bird Feeder

Brian Callanan

AT BACKYARD BIRD FEEDERS, the law of nature is sometimes eat and be eaten. The 5,000 feeder watchers who report regularly to the Cornell Laboratory of Ornithology related 946 instances during one winter in which birds seeking a free meal at feeders were attacked by predators, from raptors and snakes to cats and dogs.

The survey was done to help assess the impact of feeders on the population of overwintering birds. Dr. Greg Butcher, director of bird population studies, said, "We want to know whether feeding is increasing bird populations, or are there enough hazards at feeders to offset gains from the extra food?"

Dr. Butcher said that based on the available evidence, the good that feeders do for predatory birds outweighs whatever bad they may do for the prey. "Sharp-shinned hawks have to eat, just like chickadees do, and we have evidence that even with predation at feeders, chickadees do better with feeders than without them," he said.

Hawks, fulfilling their predatory destiny, led the pack that took advantage of the birds at feeders, according to the feeder watchers' reports. Sharp-shinned hawks alone were responsible for 26 percent of the incidents in which the predator could be identified. The Cooper's hawk chalked up 12 percent of the captures, and unidentified hawks accounted for another 7 percent.

Only cats, with 23 percent of the recorded instances of predation, rivaled the raptors' success. And Erica H. Dunn, who coordinates Project FeederWatch, an annual continentwide survey of North American feeder birds, surmises that predatory cats were overrepresented in the data because they often brought prey home to their owners, while raptors might fly off unseen with their meal.

Some hawks, however, made quite a show of their prowess. In a typical instance, described by Duff Decker of Warsaw, Illinois, a sharp-shinned hawk would sit at the edge of the woods for five to 10 minutes, then make a beeline across the yard to birds feeding on the ground or at the platform feeder.

Several hawks were seen to reach into bushes where fleeing birds had taken refuge and pull the victims out. One sharpshin pinned a bird against a window and another chased a flock of cowbirds to a lake, grabbed one and held it underwater until it drowned.

In addition to raptors, Ms. Dunn reported in *FeederWatch News*, at least 12 other bird species, including owls, jays and crows, were seen to prey on feeder visitors. All told, avian predators accounted for 77 percent of the reported incidents.

Twenty-five other species did the remaining damage, including a rattlesnake that ate cowbirds and a wood rat that caught a bird and was in turn eaten by a hawk. Dogs were credited with 11 kills, including one dog that took eight victims. The somewhat embarrassed owner wrote, "At least he only eats house sparrows."

The reported victims ranged in size from mourning doves and pigeons, which were favored by the large hawks like redtails, to small finches and sparrows. But some rather common overwintering species seemed to succumb rarely to predators. Purple finches and grackles, for example, were not often made into a meal, possibly because they tended not to concentrate at feeders or because they visited feeders only at certain times during the winter, Ms. Dunn said.

Squirrels, often the bane of people who feed birds, occasionally had their comeuppance. Two gray squirrels that intruded at feeders were seen to succumb to the talons of red-tailed hawks.

The predation count, taken during the winter of 1989–90, was not repeated this winter. However, Dr. Butcher said, "We will have to recheck predation periodically because the situation can change—what is now a plus could turn into a negative impact."

He explained: "Predation may be increasing as birds rely more and more on feeders. Five years ago it was rare to see a raptor at a feeder. Now they are regular visitors. As many as one quarter of all feeders have predators coming to visit, although most of the time these visits do not result in kills."

The reactions of feeder owners to predation is another phenomenon that intrigues Dr. Butcher. "Some take their feeders down when they see they have attracted a predator, even before any bird is killed," he noted. "Others, however, are fascinated by the bit of natural history playing at their back window. In my house, we all stop what we're doing and run to the window when a hawk comes to the feeder."

—JANE E. BRODY, March 1991

Army of Amateurs and Pros Discovers More Birds in Trouble

AT PRECISELY 4:49 A.M., on a wooded bluff above the Hudson River near Rhinebeck, New York, Otis Waterman punched the START button on a kitchen timer set for three minutes, ignored the drone of mosquitoes and listened with expert ears to the wake-up chorus of forest birds.

"Three wood thrushes," he said as their flute-like songs floated through the predawn darkness one day last week. "There's a scarlet tanager . . . a cardinal . . . a chickadee." And just before the alarm beeped, a wild turkey gobbler gobbled.

Mr. Waterman and his son Fritz, who was keeping records, piled into their car, drove a half mile and stopped by a wet meadow with a shrubby border. The habitat had changed, and so had the cast of singers, which now included a Baltimore oriole, a red-winged blackbird, a mourning dove and a common yellowthroat, a black-masked warbler with a loud voice. Mr. Waterman called the bird a "wichity," after the sound of its song.

There also were screams from a red-tailed hawk. "He's up early," Otis Waterman said, muttering something about the "ungodly hour." The protest was mild. He has been greeting dawn at this time and place every June for 28 years. A retired employee of Vassar College who lives in Poughkeepsie, he is one of 2,000 volunteers in the North American Breeding Bird Survey.

Organized in 1966 by the Fish and Wildlife Service and now part of the Interior Department's new National Biological Survey, the Breeding Bird Survey uses annual roadside counts along permanent routes to monitor population changes for more than 250 bird species, including 180 songbirds. The data can be analyzed to spot trends in particular areas like the Adirondack Mountains of New York or in broad habitat types like the northern spruce hardwood forest.

It was a 1989 review of data from the survey that alerted scientists and conservationists to an apparent general decline among neotropical migrants that breed in the woodlands of Eastern North America and winter in Central and South America and on Caribbean islands. Increasingly sophisticated analysis of the survey's numbers, however, has tempered the feeling of panic about vanishing forest songbirds. While many species show alarming declines in parts of their breeding range, it now appears that only a few have experienced major continentwide losses.

One that has is the wood thrush, a forest relative of the backyard robin whose numbers have fallen by an estimated 42 percent since 1966. The wood thrush is particularly vulnerable to parasitism by brown-headed cowbirds, which lay their eggs in the nests of other birds. In addition, the thrushes winter in tropical rain forest that is being cut pell-mell by loggers, ranchers and slash-and-burn farmers.

Another troubled migrant is the cerulean warbler, a sky-blue inhabitant of the temperate flood plain forest canopy. This species, whose population has declined by 51 percent in 28 years, is of special interest to Chandler Robbins, the Fish and Wildlife Service biologist who conceived the Breeding Bird Survey. "We've lost thousands of miles of cerulean warbler habitat to reservoirs, highways, power lines and development," he lamented. And the cerulean warbler's wintering habitat, temperate South American forest high in the Andes, is being cleared for coca plantings.

In the mid-1960's, Mr. Robbins and his colleagues laid out what they say are random routes throughout the contiguous 48 states and southern Canada and recruited skilled professional and amateur ornithologists to survey them every June, at the height of the nesting season. Each route covers 24.5 miles of secondary roads and consists of 50 stops at half-mile intervals. Starting time is 30 minutes before sunrise and the observer counts every bird heard or seen during a three-minute period at each stop. The total number of birds recorded for each species is used as an index of relative abundance.

In New York state, for example, the mean number of eastern meadowlarks counted per survey route plummeted from 20.79 birds in 1966 to 4.60 in 1993. Those numbers reflect changes in agricultural practices, both earlier mowing of hay crops, which kills young meadowlarks before they leave the nest, and the abandonment of farms where fields that have been the birds' homes are reverting to forest.

Most birds along a survey route are heard, not seen, even after the sun rises. "The observer has to know the songs and calls of every species that might be encountered," says Bruce Peterjohn, national coordinator of the survey, "and acute hearing is essential."

Noise is always a problem. Rain or wind will postpone a survey, barking dogs are a nuisance, and traffic has increased manyfold since 1966. The first stop on one route in Pennsylvania is across from a gamecock farm, and the crowing of dozens of roosters at daybreak drowns out the bird song.

Nor is every route run with the clockwork timing that the survey office prefers. Surveys have been stalled by a major earthquake, herds of sheep, a loose and friendly pig, suspicious police officers, a house being moved that collapsed on the road, even a marching band. An ornithologist whose urban route begins near a steel plant in Gary, Indiana, had to ward off prostitutes waiting for a shift change. One volunteer drove into what appeared to be a shallow stream in Colorado and was stranded in deep water for several hours while spadefoot toads swam around his car and serenaded him with appropriately duck-like quacks.

There were no unusual incidents this year on Mr. Waterman's survey route, which twists and turns across the farmland of northern Dutchess County. But there have been significant changes since 1966. Fields that once sprouted corn are overgrown with housing developments. The briary thicket where in 1966 he discovered a yellow-breasted chat, a rare breeding bird in New York state, is now the first tee of a golf course, and the only birds in sight or sound are the ubiquitous Canada geese on the clubhouse pond.

The Breeding Bird Survey was designed to monitor the impact on breeding bird populations of such subtle land-use changes as well as catastrophic habitat destruction by forest clear-cuts, mining and suburban sprawl. No one, however, anticipated the degree of controversy that was set off by the 1989 report on the decline of neotropical migrant songbirds, written by Mr. Robbins and three colleagues.

"I'm high on the Breeding Bird Survey," said Dr. Frances James, a biologist at Florida State University in Tallahassee. "It's our best source of information. But the data do not show big declines in neotropical migrants except in a few cases. Half of the species are increasing, half are decreasing, which is what you'd expect if nothing major is going on."

She pointed out that the solitary vireo, a bird that prefers mixed wood-lands, has been increasing in Eastern forests at a remarkable rate of 4.2 per-cent a year. "That doesn't mean we shouldn't be studying neotropical migrants, finding out which birds are in trouble and helping them out," she said.

Dr. James expressed special concern for the prairie warbler, a bird that once was abundant in Southeastern scrublands and has declined by 45 per-cent in its range since 1966. "Our analysis of B.B.S. data also shows that species which are increasing in general are declining in highland areas like the Adirondacks, Blue Ridge Mountains and Cumberland Plateau," she said.

Peter Stangel of the National Fish and Wildlife Foundation in Wash-ington said, "The fact that a species may not be declining rangewide should not stop us from starting conservation programs." Spurred by the Robbins paper and other disquieting reports, the foundation began a campaign that brought together federal and state wildlife agencies, private conservation organizations, scientists and the forest products industry in a Neotropical Migratory Bird Conservation Program.

"Too often," Mr. Stangel said, "we wait for disaster before taking action."

No one disagrees with reports based on Breeding Bird Survey data showing that many prairie songbirds like the grasshopper sparrow and lark bunting are in serious trouble. Most of the species are short-distance migrants and never leave North America. Loss of grassland habitat may be the major cause of their decline.

There is growing concern, too, for shrubland birds like the rufous-sided towhee, whose familiar spring song is interpreted as "drink your tea." As natural plant succession turns thickets and brushy fields into forests, the towhee loses ground. The mean number of towhees counted on survey routes in New York fell from 10.86 in 1966 to 2.43 in 1993.

Mr. Waterman's Dutchess County survey reflects that precipitous decline. In 1966 he logged 23 towhees. Only seven were heard this spring.

—LES LINE, June 1994

A Lone Naturalist Defends
Secret Places of the Falklands

THE EXTREMITIES of earth can be a naturalist's paradise. And the Falkland Islands, this barren grouping of islands in the South Atlantic 800 miles north of Antartica and 300 miles east of Argentina, is such a place.

Consider the population: two million penguins, up to two million albatrosses, millions of petrels, 40,000 seals, as well as some of the rarest birds in the world, including the upland goose, the flightless steamer duck and the mischievous striated caracara, a falcon that one Nantucket sealer in the early 1900's described as ever pilfering "caps, mittens, stockings, powder horns, knives, steels, tin pots, in fact everything which their great strength is equal to."

Indeed, the bird life is so abundant in some parts that the islands' government issues a map carefully detailing areas for pilots of small planes to avoid.

But the presence of man and sheep over the last 150 years has slowly pushed the vast population of wildlife here to the islands' own extremities. Species of birds once common on the two main islands are now found only in some of the Falklands' outermost islands and islets.

And even on several of the offshore islands, the struggle between the demand for bigger flocks of sheep, currently around 750,000 head, and the natural ecological balance continues.

The Falklands' one conservationist notes these trends with concern.

"We've messed up the main islands and some of the offshore islands have been hit pretty hard," said the naturalist, Ian J. Strange, who has been studying and documenting wildlife here for 31 years. "These islands are very small and just a few sheep can create a big disaster."

The islands are not idyllic for human life (currently 2,100 people inhabit the Falklands) but they seem to be for bird life. And a brief review gives the impression that the Falklands could be one of the greatest con-centrations of bird life available for study. The main reason is an extremely large food supply created by ocean currents.

As the cold waters of the southern Antarctic Ocean move around Cape Horn toward this outermost British colony, they split into two different cur-rents, one moving to the south of the islands and one to the north.

As the currents reach the islands, underwater ridges reduce the normal depths and create strong tidal actions and upwellings, creating a natural fun-nel for huge amounts of food, especially krill and squid. These immense reserves of food have helped sustain vast colonies of birds.

Mr. Strange, who is self-employed and supports himself by selling wildlife illustrations, has documented islands where tens of thousands of rockhopper penguins and black-browed albatrosses nest intermingled within an area half the size of a football field.

A principal reason why wildlife has been pushed to the outer islands, Mr. Strange notes, is the almost total destruction of the native tussock grass that was once the dominant vegetation. This grass, which grows 12 feet high at times, is unique as a shelter for bird life, like wrens, thrushes, short-eared owls, vultures and the striated caracara.

Once sheep farming was introduced, the grass's disappearance was inevitable. Now tussock, and much of the bird life, can be found only in places where there are few or no sheep, like the outer islands.

New threats to wildlife have developed from the economic prosperity that followed the unsuccessful 1982 invasion by Argentina, which claims sovereignty over the Falklands. Hundreds of boats a year now ply the waters harvesting the vast reserves of squid around the islands, an industry that since 1988 has brought in as much as $50 million a year.

Mr. Strange said that on one isolated island he has studied, near the center of squid fishing, the rockhopper population has fallen by one third over the last three years.

"Some serious thing happened to that colony, and the only thing I can say is that it must be related to the reduction in the availability of squid in the area," he said.

Sea lions have also suffered. There were up to 630,000 in the 1930's, but Mr. Strange found only 63,000 in a survey in 1960 and a mere 8,000 in 1990.

"I knew they were still on the decline, but when I found out what really was happening I was shocked to the core," he said. Yet he has no clue as to the cause.

All these developments have moved Mr. Strange to mount a campaign to protect as many islands as possible from further commercial development. Starting in 1962, he began prodding the Falklands Legislative Council to designate specific islands as nature reserves and has since won such status for 50 islands. But the council's acts are not irrevocable and future governments could revoke them and permit development.

So Mr. Strange has also urged islanders to buy individual islands and create irrevocable trusts that preserve them as wildlife sanctuaries. In 1971, along with a partner, he bought New Island, one of the westernmost islands of the Falklands, and set up a wildlife preserve to study the colony of 3,000 fur seals, 10,000 albatrosses, 100,000 rockhopper penguins, 10,000 gentoo penguins and an unknown number of burrowing petrels called thin-billed prions. The island has one of the largest known breeding colonies in the world.

But since then he has struggled to raise even a fraction of the $1.2 million needed to set up an endowment.

Such actions have not all gone unopposed. Although he has lived on the Falklands for more than three decades, he says he still feels like an outsider among the native islanders, who he says are suspicious of his efforts to create wildlife preserves.

One issue he has faced several times is whether to make public the whereabouts of islands he has discovered that are so pristine that the birds have no fear of humans. In these places he has been able to walk among the wildlife and have birds hop on his boots and caracaras steal lunch from his hands.

Several years ago, in what he calls one of the biggest mistakes of his life, he published accounts of his trips to the most southern of the Falkland islands, Beauchene. Lying 100 miles south of the East Island, Beauchene was so remote that until he first stepped on it in 1963, he believes no human had been there for at least 40 years, and perhaps never.

There, the dense colonies of albatrosses and penguins were breathtaking. Tens of thousands of petrels and prions crowded the islands at night after spending the day gliding over the ocean. He also found the largest concentration of striated caracaras.

It was here that he was able to do his most extensive work on the caracara and study the ecosystem of the tussock grass.

But once word began to leak out about the location of the island and its uniqueness, others wanted to visit. And he began arguing that such places should be left alone, even by scientists, except for very infrequent visits.

So far he has won only half of the battle. Visitors can go, but only with the permission of the Falklands government, and no one has won a permit for over a year. Still, he said he was concerned that the decision-making mechanism was not more structured.

"If you try to count the number of really pristine areas left in the world, they are very, very few," Mr. Strange said. "They should be left entirely alone and used as yardsticks."

Mr. Strange argues that such places, where the wildlife has never seen a human, should be studied only once every four or five years, and then only by aerial observation. If scientists must enter such islands, he says, they should not be allowed to stay for more than a few hours at a time.

The battle over Beauchene left an indelible mark on the 55-year-old naturalist, and he plans to carry several other secrets about the Falklands to the grave with him, in an effort to protect the islands' wildlife from man's intrusion.

He says he knows of three other sites on the islands, similar to Beauchene, where nature is pristine and man is virtually a stranger. And he will not disclose the locations.

Also, while most naturalists say there are five types of penguins that nest on the islands—the rockhopper, the gentoo, the magellanic, the macaroni and the king penguin—Mr. Strange says he has discovered a sixth.

Somewhere among the Falklands' 200 islands and 150 islets is a colony of royal penguins, its location known only to Mr. Strange.

"It's an issue of simply protecting these penguins from the intrusion of man and I think that's important," he said. "The environment here is very, very fragile. It's really little more than a small garden. A horse gets in and he can trample it to death in a very short time. It's not like a large continent where there is a buffer effect."

—NATHANIEL NASH, July 1991